Prison Journey

A Memoir

Prison Journey

A Memoir

By

Brigadier F. B. Ali

V

VANGUARD BOOKS

ISBN: 978-969-402-573-5

Plot # 52-53, Gulberg-II,
Main Guru Mangat Road, Lahore.
Ph: # 92+42+35764792 Fax: # 92+42+35764794
email: info@vanguardbooks.com
www.vanguardbooks.com

Printed at:
Banquet Printing Press
Lane # 28-A, Nisbat Road, Lahore
Phone No. 92-42-37245590

For Jamila

My companion on this journey, without whose selfless support, steadfast courage and determined efforts on my behalf, I would not have survived it.

And for Afroze and Yawar

Who, without being asked, got taken along on the ride but maintained their good humour, coping with the ups and downs of the journey, and joined in the fight to bring it to a satisfactory conclusion.

Also for 'the band of brothers' of Attock

Who, for the sake of their country, dared to look beyond the narrow bounds of duty and sacrificed their liberty and their futures in the endeavour, and then cheerfully bore their adversity with fortitude and great courage.

کچھ لغزشوں سے کام جہاں کے سنور گئے

کچھ جرآتیں حیات پر الزام بن گئیں

(باقی)

Sometimes the coward's trembling
Will right the world's wrongs,
Sometimes the daring of the brave
Remains just a silent reproach
At the fickleness of Fate.

– Baqi

Contents

Pictures

Preface

The decade beginning in 1969 was the start of Pakistan's time of troubles. Caught up in the events that marked this period, my attempts to do what I could to stop the downward spiral were unsuccessful, and finally resulted in my spending more than 5 years in prison, most of it in solitary confinement.

From time to time family and friends have urged me to write about those times and my role in what transpired during them. But I found myself reluctant to revisit a period that was for me a painful memory of loss and failure. However, what finally tipped me over the edge was when the issue was put to me in terms of duty. My nephew Omar told me it was something I owed to the younger generations of the family; my friend Mehmud Ahmed said I owed it to history.

In writing about that time in my life and in the history of Pakistan, I have based my account on the various documents that I had, especially the journal that I wrote while in prison. Many readers will not be familiar with the events of that time, so the first part of this book deals with the road that led me to that first prison, the fort at Attock.

The second part covers the time I spent in various prisons in the Punjab, and is based on my prison journal. Wherever appropriate, I have quoted from the original account, or have used it as the basis for the narrative. As additional context, I have provided in the Introduction a short factual narrative covering some of my background and career in the army.

Now that it is done, I am glad I was pushed into writing this account of part of my life and times, and grateful to those who did that. I am greatly beholden to my friend, Ingolf, for all his assistance and wise counsel in this undertaking, including reading through several successive drafts of the manuscript and helping to improve it. But even more so for his friendship these many years; it is a rare gift to discover a mind and spirit so in tune with one's own, especially at this stage of one's life when Time has robbed one of the few real friends made during life's journey, leaving a large gap in one's life.

I am deeply thankful to all those – relations, friends, acquaintances, and even strangers – who, by their support, help and acts of kindness, enabled me and my loved ones to cope successfully with this difficult time in our lives.

Introduction

My father was the son of a small-time farmer in the princely state of Patiala (in what was then called British India). He was orphaned at a young age and was brought up by his older brothers. Through sheer ability and application he completed his education by obtaining a Master's degree from Government College, Lahore, and then won a place in the Indian Police service through an all-India competitive examination. My mother had Central Asian ancestry, her mother having come to India as a young girl from Turkmenistan while her father was a Lodi, a tribe that had originally come to India from Afghanistan.

I was born on the 20th of February, 1929, in New Delhi, where my father was posted as a young police officer. My brother, Kaiser, was born a little over a year later. My first connected memories of childhood are of living in Karnal, where my father was posted as the superintendent of police; I must have been about 4 years old. One of these memories is of our house being wired up to receive electricity from the first power station set up in the town.

My parents became concerned about the adequacy of the local schools and the likelihood this would also be the case in the other places to which my father was likely to be posted every 3 years or so. After some research, when I was about seven, my brother and I were placed in a boarding school in Mussoorie, up in the hills, which had a good reputation – the Convent of Jesus and Mary. We knew practically no English, which is what was spoken at the school, and we had never had to manage on our own. The first few months were hard; I grew up in a hurry (I suppose my brother did, too). After this initial rough period I began to enjoy the school-life, though I am sure I must often have been homesick that first year.

We spent 4 years in Mussoorie, finally leaving the school in late 1939. Our father was then superintendent of police in Hoshiarpur, and we joined schools there; I took my matriculation examination in 1942. Soon after, my father was transferred to Muzaffargarh but, after a short stay there, I left to go to Lahore where I had obtained admission into the Government College, my father's alma mater and considered the premier educational institution in the province.

My first two years there were a lot of fun. Living independently, especially in a big vibrant city like Lahore, was exciting and enjoyable. As was attending classes taught by good professors in the ornate old college buildings, and the communal life in the hostel. The college had an excellent library and I made good use of it, spending a lot of time reading. In sports, I took up boxing, and became quite good at it.

I used to spend the summer vacations at home in Muzaffargarh, during which I would accompany my parents during my father's frequent tours of the district. We would stay in small rest houses in the countryside, often situated on the banks of canals, with wide open spaces all around dotted with fields and trees and small villages. I got to see and learn something about the ordinary people who lived in these villages, simple folk who lived hard lives of continuous toil, gaining an intuitive feeling for my land and its people. There was also the desert; our house was on the edge of town and the sand dunes began about a mile away. I took long solitary walks, thinking about my future and the world into which I was growing up.

These experiences and reflections had their full impact when I returned to college in my third year there. I was no longer the carefree young person enjoying college life, as my fellow students were, but instead became acutely conscious of the times I was living in, and the need for me to take part in the struggle of my people. The time was 1945, and the Muslims of India were locked in a bitter struggle against the British and the Hindu Congress (and some of their own compatriots) to achieve a homeland of their own – Pakistan.

Muslim students were taking an active part in this movement – except in Government College, Lahore. Like students in general in this institution, and in line with its traditions, they showed no interest in political activism. I decided to remedy this. It took me many months of continuous effort but, with the assistance of some others, I managed to set up the first unit of the Muslim Students Federation in the college and enrol the majority of Muslim students as members. I was elected its President, and later became a member of the Working Committee of the Punjab Federation. (Some years ago, visiting the Pakistan gallery of the Lahore Museum, I was pleasantly surprised to see displayed there a picture of this Working Committee with the founder of Pakistan, QA Muhammad Ali Jinnah).

These two years were a whirlwind of activity, mobilizing the students of my college to take part in all the political activity going on – public meetings, processions, demonstrations; sending groups of students into the countryside to canvas for Muslim League candidates during elections; once being in-charge of the contingent of students guarding Mr Jinnah during his overnight stay at Mamdot House (the Khaksars had threatened to assassinate him). Inevitably, my studies suffered; I did not care but my parents were concerned, though they were very understanding. In 1946 I appeared in the BA examination, and passed with average marks. During this period I got to know two other young activists who became lifelong friends, Riaz Piracha (who later joined the Pakistan Foreign Service) and Qasim Rizavi, a prominent student leader (who ultimately joined the Civil Service of Pakistan).

I was only 17 – too young to think about getting a job. I joined MA classes, but soon dropped out; I could not summon up enough interest. Things were getting ugly: with the British bumbling along trying to devise some acceptable post-Raj setup in India, the various communities were jockeying for position, and riots and killings were breaking out in different places. It seemed to me these clashes would widen, but the Muslims were not prepared for such a development. I wrote a proposal to remedy this by training and arming a corps of students and young men, and discussed it with Qasim Rizavi. He took me to see Mumtaz Daultana, a Muslim League leader, who told us he would raise the matter in a top level Muslim League meeting that he was attending in Delhi in a couple of days.

When Daultana returned from Delhi he told me he had given my proposal to Khurshid Anwar, who had been appointed to organize the Muslim League National Guards, and I should get in touch with him. Khurshid Anwar asked me to organize the student wing of the Guard, putting me in charge of it, but events were moving too fast for either of us to do much organizing. Khurshid Anwar had a bent for 'cloak and dagger' stuff, and I became involved in some of his ventures. This was a bleak period for me; I was living in pretty Spartan style, sharing a room over the detached garage of a League bigwig's house with another operative, frustrated at the lack of a clear path ahead, with little to show for my time and efforts, and the situation in the country getting worse by the day.

Meanwhile, Khurshid Anwar had got involved in other things. In the Punjab the Muslims had launched a protest movement against the Unionist party government, and it was in danger of fizzling out because the leaders were all being arrested; he went 'underground' and sustained and directed the movement until it succeeded in forcing the government to call fresh elections. He then went off to the NWF Province to repeat that tactic against the Congress government there. In August 1947 when I met him in Karachi, where we both had gone to witness the birth of Pakistan, he told me that he had been given the go-ahead (I guessed by Liaquat Ali Khan, who was becoming the Prime Minister) to lead a force of tribesmen into Kashmir, an adjoining princely state, to ensure its accession to Pakistan, and asked if I would like to accompany him. I readily agreed, but on my return home my father put his foot down and vetoed the idea.

I then decided to join the Pakistan Army, and was duly selected. In January 1948 I reported to the newly established Pakistan Military Academy at Kakul, already housing the Muslim cadets from the Indian Military Academy who still had one year of their course remaining. The time at PMA was both busy and tough, but I enjoyed the challenge and the opportunity to learn my new profession. In the hectic activity and easy camaraderie of our training I became friends with several course-mates, and developed a lasting friendship with one of them: Ateeq Ahmed. We remained close till his death a few years ago.

Our one-year Graduates course was cut short due to the fighting in Kashmir, and we passed out in November along with the IMA-PMA course. I had wanted to be commissioned into the Armoured Corps (based on the influence of the military writers I had read, especially Maj Gen JFC Fuller) but, even though topping my course entitled me to the only vacancy available, someone else used their influence to grab it and I was posted to the Artillery. Later on, I was very happy this had happened when I discovered that the only two combat arms in the army that were truly 'professional' were the Artillery and the Engineers.

In June 1952 my brother Kaiser died. He had joined the Navy and had gone to England on a long training course; there he was involved in a fatal collision while driving to London. It was a devastating blow to my parents. Their grief was somewhat blunted by having a young son to look after. Daniyal had been born in 1947, and became a source of great joy to our parents, filling some of the emptiness in their lives when Kaiser and I left home. Kaiser's loss was a great tragedy; he was a late bloomer, but had grown into a very smart and accomplished young man.

Later that year I went on an instructor's course to the School of Artillery, and was then kept on there as an instructor. In November 1953 Jamila and I were married (something we had both set our hearts upon many years earlier). Life in Nowshera as a young newly married couple was idyllic, with several other young couples there who became lifelong friends. Our daughter, Afroze, was born in early 1955. She was only a few months old when we went to England for a year's course at the School of Artillery there. This, too, was a wonderful experience. My father had retired and the three of them came and stayed for some months with us. We took advantage of their presence to leave Afroze with them while we drove around Europe during the summer holidays.

After the course I returned to the School of Artillery as an instructor. The army was undergoing a major change as American equipment began to flow in under a military aid agreement. I was put in charge of a small group of instructors to develop new artillery doctrine and procedures that could apply to both the new US weapons and the old British ones we already had.

Snapshots

Friends since college days: with Riaz Piracha (middle) and Qasim Rizavi, 1949

Wedding picture (1953)

The family, 1958 (*L to R*: My father, Daniyal (my brother), Afroze, my mother, Jamila)

Our family, 1965

This was a very fulfilling time; we developed and produced a whole set of new training manuals for the Artillery, and ran courses to train the other instructors in these new methods and procedures.

During his visit to us in England my father had suffered a heart attack. Back home he was not keeping good health. I asked for and got a posting to Lahore so that we were able to live with him and the family and it was here in 1959 that our son Yawar was born. I had taken the entrance examination for the Staff College, and was selected to attend a foreign staff course, a prized outcome. However, my father's condition was not improving, and I declined the course. In 1960 my father suffered a fatal heart attack. I reappeared for the staff course examination and was selected to attend the Canadian Army staff college in 1961. The two years we spent at Kingston were an exhilarating time for both us and the children; we made many friends and developed warm feelings for the country and its people.

On our return in 1963 I was posted as second-in-command of a medium artillery regiment in Lahore. A few months before the 1965 war broke out I was called to GHQ to join a newly created Research & Development Directorate. Immediately after the short war I was posted as second-in-command of the newly raised 44 (SP) Field Regiment, but after a few months was promoted Lt Col and called back to the R & D Dte in GHQ.

From 1967 I commanded 44 (SP) for two years, and was then promoted in early 1970 to chief of staff of 18 Division in Karachi. I was promoted Brigadier in early 1971 and became Deputy Commandant of the Pakistan Military Academy; shortly before the 1971 war I was ordered to take command of the artillery of 6 Armoured Division. I commanded this formation during the war, but shortly thereafter was posted as Commandant of the School of Artillery.

In August 1972 I was compulsorily retired from the army.

Soldiering

Pakistan Military Academy (25 November 1948)
(Receiving the Academic Gold Medal from the Governor-General, Kh Nazimiddin)

With the Engineers (School of Infantry & Tactics, 1949)
(*L to R*: Salamat, Shafiq, Me, Anwar)

School of Artillery, Larkhill (UK), Long Gunnery Staff Course (1957)

School of Artillery, Nowshera (1958)

Canadian Army Staff College, Winter 1962: Foreign students camping in the 'bush', Camp Gagetown (*L to R*: Me, Marino Lucarelli, Ove Pederson, Nimmi Puri)

Research & Development Directorate, GHQ (1966)
(With (then) Brig Shaukat Riza)

44 (SP) Field Regiment, Artillery (1969)

Full Circle: Pakistan Military Academy, 28 August 1971 (Presenting the Sword of Honour to Bn SUO Nadeem Ahmad Aziz)

PART 1

THE ROAD TO ATTOCK

1

The Beginning

Where did it begin, this road that took me to the sprawling stone fort of Attock, perched on the low hills above the winding Indus River? And then, from there, through the shadow of the gallows, to those many prisons of the Punjab before I finally ended up here in Canada?

Confucius is reputed to have said: *The journey of a thousand miles begins with a single step*. What, then, was this first step that set me upon this long road that I have travelled?

Was it the sudden ending of my childhood when, at the age of seven, I was plucked from the warm cocoon of a small family and deposited in the midst of a boarding school up in the hills? I had never been away from home, had never had to fend for myself, did not know any English, which is what they spoke there. I had to grow up fast, very fast.

Or, was it all those new and exciting ideas of nationalism and freedom from foreign rule that Taufiq put into my innocent head? My friend, my difficult friend (I was the only one he had in that school), whose precocious mind knew much more about the outside world than I did, and teemed with ideas that I knew nothing about. I still remember the excitement of those "rambles" on which we were occasionally taken out, when, while the other boys played boisterous games around us, I would sit, rapt and entranced, on the sun dappled slope of the wooded hill while he told me the tale of the Count of Monte Cristo, or held forth on how India had to overthrow its subjection to the British.

Poor Taufiq! He could not cope with a world that neither understood nor cared for the brilliance of his mind, the ebullient joy of his spirit, the essential innocence of his soul, a world that would make no allowances for the peculiarities of his nature, a nature that he had neither fashioned nor sought. Finally, unable (or unwilling) to carry on this constant struggle, he went to a railway platform and stepped off it, departing from a world that seemed to have no use for him.

Was it that intoxicating time when I was consumed by the passionate struggle to create Pakistan? My first two years in college had been spent in the heedless but innocent enjoyment of living away from home in a big city. Then, suddenly (it was almost like a religious conversion) this struggle was the only thing that now mattered in my life, and I threw myself into it completely: organizing my fellow students, marching in the processions that mobilized citizens, facing off the police and hostile activists, trudging through the countryside, village after village, mile upon weary mile, seeking votes. Then, as ominous clouds began to gather in the sky, engaging in cloak and dagger escapades in the murky, shadowy world of Khurshid Anwar, risking much more than I then realized.

With Pakistan achieved, I abandoned any thoughts of seeking a potentially lucrative career and, instead, joined the new country's army — having had a hand in its creation, it seemed natural to now help in defending it. Soon, too soon, its founder's death sent the new state into a spiral of regression; as I watched it lurch from one crisis to another, its nascent democracy abandoned, my service in the army became for me a path to a position from where I could, one day, do something to restore it to what it was meant to be. This secret goal, nurtured in my heart, lightened the monotony of peacetime soldiering, and the inevitable rough patches one encounters during it.

Surprisingly, there came a time when it seemed to me no longer necessary to pursue my secret goal. In 1968 the political and public agitation against President Ayub Khan's rule forced him to concede to the demands for a return to the parliamentary system of government, with elections to be held in 1969. What really gave me hope was that a number of patriotic and honest leaders, most notably Asghar Khan (the former chief of the Air Force), had led the movement for the restoration of democracy, and looked set to win these elections. I seriously contemplated leaving the army and working with Asghar Khan to help him in putting Pakistan back on the right path.

All these hopes came crashing down on that fateful day in March of 1969 when Tony suddenly and unexpectedly appeared at our house in Kharian cantonment. That certainly was the first step in the final portion of the road to Attock. The news he brought sickened me, and set me inevitably upon the path that led me to that grim fortress prison, and all that came after.

The real beginning of this journey, though, goes further back, much further back. The path I took was written into my being the day I was born my father's son. What he bequeathed to me made it inevitable; with that heritage I could have followed no other path. (I make an attempt to depict something of the person he was in Appendix 1 (page 46).

That I survived this hard journey, more or less intact, I also owe to my parents. My father had the strength of tempered steel: you could bend it to excruciating extremes, but it would not break. My mother had a core with the hardness of granite: it would chip, but it would not shatter.

2

Days of Darkness

1968 had been a time of much turmoil in Pakistan. Fed up with his ten years of presidential rule, the people had mounted a sustained agitation against Ayub Khan. Finally, unable to quell it, he promised to allow a new parliamentary constitution to be drawn up, followed by elections. A new era of democracy and renewal promised to dawn soon. There were grounds for hope that Pakistan could once again resume the path that its founder had wanted for it, the goal so many of us had struggled and fought for.

It all turned to dust and ashes for me on that evening in March 1969. We were living in Kharian and I had taken a few days off to visit my mother in Lahore. Our bags were loaded in the car and we were about to get in when a jeep turned into our driveway. I was very surprised to see my friend Tony[1] emerge from it, since he was supposed to be in Multan, serving in the Corps HQ there. Drawing me aside, he told me that he had just delivered to my division commander the orders for the imposition of martial law in a few days. I was devastated. As we were ready to leave, I arranged to meet him later that night in Lahore.

---

[1]Lt Col (later Brigadier) Mir Ijaz Mahmood SJ. We had passed out together from the Military Academy; later we became good friends.

After dinner I met him at the Punjab Gymkhana Club; we sat outside in my car and talked. I told him it had to be stopped somehow, though I did not know how. As commander of an artillery regiment, there wasn't much I could do; I hoped Tony might have some ideas. But he thought the imposition of martial law wouldn't be a bad thing, and I could see he didn't understand my agitation at the prospect. We parted with his attempted reassurances echoing in my ears.

I sat up that whole night, racking my brains to come up with some way out, agonizing over the dead ends I kept running into. Jamila would wake up at intervals, see me still sitting there, and keep suggesting I go to bed. Finally, I came up with the only thing I could do which might avert this catastrophe: I would get word to Asghar Khan. Perhaps he would be able to do something about it.

I got ready and, as dawn broke, drove to Manzar Bashir's house. I had known him in college when we were both active in student politics; he was now a senior officeholder in Asghar Khan's political party. His servants had to rouse him from his sleep, and he came down in his dressing gown. I told him what was going to happen. He said Asghar Khan was in Karachi, but he would fly there immediately and brief him. That evening I saw him again after his return. He had managed to get a few minutes with Asghar Khan between his meetings; when he told him, the reply he got was: *If they are stupid enough to do that, there's nothing I can do about it.*

So, a few days later, that black period of shame descended upon the land. For me the way to live through it was to nurture in my heart once again the secret mission of waiting and preparing for the day when I would be in a position to do something to set things right. Meanwhile, throwing myself into soldiering offered a way of doing something useful while ignoring the filth that was slowly building up outside.

But, sometimes, the dirt seeped into our cantonments, too. One day the commander of the local military hospital told me how, some days earlier, the infamous 'General' Rani (the small-town madam who had achieved great influence by supplying girls to President Yahya Khan and his coterie) had arrived at his hospital wanting to be admitted into the VIP ward; he had naturally refused. A couple of days later he was woken up in the middle of the night by a phone call; at the other end was Yahya himself, obviously drunk, who gave him a dressing down for not accommodating Rani.

I experienced some of this myself when, in the summer of 1969, I spent a few months as martial law administrator in Lyallpur. One day a man came to see me bearing a letter from Lt Gen Atiqur Rehman, corps commander and martial law administrator for West Pakistan, in which he asked me to help the bearer with his problem. Having read the letter, I looked at the man who had brought it. He was short, fat and slovenly, with a few days' stubble on his face, shifty-eyed, loose-lipped. He said his brother owned a factory in Lyallpur; he wanted me to take it from the brother and give it to him!

I looked at the man, and then I looked at the letter. Atiqur Rehman was the chief instructor at the Military Academy when I was there. He was considered a very capable professional soldier, the epitome of all the soldierly virtues that we were being taught as cadets. That he had come to this saddened me greatly. I dismissed the man, saying I would look into the matter. Enquiries through my police liaison confirmed that the brother was indeed a factory owner, a thoroughly respectable citizen who had some years ago thrown out this dissolute brother of his. I gave orders that he was not to be allowed into our offices again.

A few days later, I was woken up in the middle of the night by a telephone call from my division's chief of staff. He said that President Yahya Khan had just called the division commander asking why this man sent to me had not been helped, and that the general was very angry about it. I replied that there were no grounds to do what the man wanted. Next day I received orders not to deal further with this matter since an officer was being sent from Kharian for this purpose. I learnt later that this colonel had the factory owner arrested on some trumped up charge, and then forced him to give part ownership of the factory to his brother. It seems that the latter was a protégé of 'General' Rani, and she pushed a drunken Yayha into helping him.

With the tragic was mixed the farcical. I was staying in the Circuit House in Lyallpur. One day Maj Gen Khuda Dad, the army's Adjutant General, arrived for a night's stay there. He held a meeting in the evening, and then departed for dinner with the manager of the local Standard Bank (this was the bank patronized by the generals; all the army's funds were required to be deposited in it; in return, it was rumoured, generals got 'loans' to build houses and for other purposes). Later that night there was a knock on the door of my room; it was the general. *Come,* he said, *I want to show you something*. He took me to his bedroom where, laid out on the bed, were a dozen felt hats in different colours and styles; this was his 'loot' of the day. I mumbled some words of admiration and beat a hasty retreat.

The work of 'administering' martial law in the district was supposed to be general oversight of the civil administration, but actually consisted mainly of dealing with an endless mass of complaints that people brought to us on matters where they felt they had been wronged or mistreated by that administration or other people. Towards the end of my stay in Lyallpur I wrote a letter to my old friend Riaz[2] which expresses the thoughts and emotions that this experience engendered in me:

"This has been a strange experience. I find that, for the first time since 1947, I have had a chance to see how the bulk of our people really live. We move in such a narrow circle, a sanitized bubble of glass, completely cut off from the life of the people, the smells, the squalor, the shrieks, the sheer grinding misery of it all. How shall I describe it to you? How can one describe a nightmare?

The closest simile is that of a jungle. In this jungle are certain protected enclaves, in which the fortunate few, the 'untouchables', live (our good fortune and our shame is that we are among these). The vast majority of the population lives in the jungle outside these havens. The wardens of the jungle are the government officials. There is no law in the jungle, only the whims of the wardens. Everyone in the jungle is fair game. The wardens hunt for pleasure or for profit. They license some of the protected to hunt, too. Everyone is fair game. You can kill, you can crucify. You can bleed them to death, you can suck the marrow from their bones. You can rape, you can ravish. There is no law in the jungle save the wishes of the wardens.

Within this night of terror lie the warm, brightly lit islands of the protected, the untouchables, us. The soft hum of the air conditioner provides a smooth backdrop for the Mozart pouring from the radiogram. The laughter of our bright-eyed, well-scrubbed children echoes the precocious chatter of Dennis on TV. We talk of art and literature, even of economics and politics. We talk of plans, reforms, the

---

[2] Riaz Piracha was then Ambassador to Belgium and the EU; he ultimately retired as the Foreign Secretary.

future, progress. We discuss, we debate, we feel 'involved'. But these are all fantasies, hallucinations produced by the drug we take to shut out the terrible sounds that rise from beyond the walls. Without this drug one could not retain one's sanity.

I was thrust, unprepared, into this jungle. The hunting is a little subdued now, the killings are done in the dark, the rapes take place in the thickets. The moment we go back to our cantonments, it will revert to normal. Nothing has really changed, nothing can change.

What can a man do? Confronted by this mountain, this veritable Himalaya of injustice, misery and horror, how can you start to dismantle it with nothing but your bare hands?

What can one do? What, except turn away shuddering, and pop yet another pill of forgetfulness into your mouth. If one life could right all these wrongs, I would gladly give mine, but that is not the way things work.

You can forget, you can turn away, but how can you shrug off the crushing burden of helplessness, of grief, of guilt?

And all the time, beneath your feet, you can feel the steady throb of the irresistible force of Life, blind to all else, just producing, adding, multiplying — children, victims, ...avengers?

We may escape the reckoning, but I fear greatly for our children".

****

Early in 1970 I was promoted to Colonel and posted as chief of staff in the division HQ in Karachi. The division commander, Maj Gen M Rahim Khan, was also the martial law administrator for Sind and much preferred those duties, leaving the running of the division largely to me. This suited me fine, and I plunged into the work involved, especially the development of a suitable tactical concept to defend the large desert border we were responsible for (surprisingly, there was none; the same doctrine used in the plains of the Punjab was being applied here in the desert).

However, I could not remain completely isolated from the less savoury goings-on around me. The major political event for that year was the general election — the first in Pakistan's history — to be held towards the end of the year. The military regime's main aim was to ensure that no political party emerged in either wing of the country (especially East Pakistan) with a decisive majority, so that it could continue to manipulate the political process to retain power. To achieve this they provided support to certain political parties, the main one being the Jamaat Islami in East Pakistan (as a counter to the Awami League). One of the ways of doing this was to funnel large sums of money to them by forcing rich businessmen to make contributions.

Once, Gen Rahim, who had gone to Rawalpindi to attend some meetings, phoned me to arrange for Seth Dawood (one of the richest tycoons of Pakistan) to see Maj Gen Farman Ali Khan next day at the division HQ. Farman was an advisor to the governor of East Pakistan, and was heavily involved in the regime's games there. Farman arrived the next day from Pindi (on his way to Dhaka) and, as arranged, Seth Dawood came to see him. After the meeting the Seth came into my office, looking rather pale; I could imagine the arm-twisting that Farman must have indulged in to force him to contribute money to his political slush fund. Saeed, a PMA course-mate of mine, was involved in these activities in East Pakistan, and once told me how he used to travel around with briefcases full of money.

In spite of all this, the elections (which were not otherwise manipulated) were a disaster for the regime: Mujibur Rahman's Awami League won a big majority in East Pakistan while Zulfikar Ali Bhutto's People's Party did the same in the West wing. As a consequence, Gen Rahim, who was in the inner circles of the regime, now started wooing Bhutto, who visited him in our headquarters several times. These meetings obviously bore fruit, and some kind of understanding was arrived at between the military regime and Bhutto. This resulted in the duplicitous dealings of the regime with Sheikh Mujibur Rahman, and ultimately led to the military action in East Pakistan.

As the situation deteriorated, I became increasingly disheartened at the direction in which matters were heading; I could see the coming disaster. Adding to my distress was the jingoistic attitude being displayed by most West Pakistanis, both military and civilian. People whom I regarded as sane and balanced openly talked about suppressing the Bengalis by force; there was hardly anyone who shared my view that this policy was both morally wrong and practically ruinous. Two notable exceptions were Lt Gen Sahibzada Yaqub Khan and Vice Admiral SM Ahsan. The former was sent to East Pakistan as Commander Eastern Command and later became governor of the province. In both capacities he urged Yahya Khan to negotiate a settlement there; when Yahya decided, instead, to solve the problem by force Yaqub Khan resigned and stoically accepted the unpleasant consequences that followed. Admiral Ahsan preceded Yaqub as governor and also resigned in disagreement with the regime's policy there.

Unable to do anything to influence the course of events, and sick at heart, I decided to leave the army. But I had a family to support, so I set about finding some alternative employment before resigning. I talked to some friends, and they arranged meetings with a couple of prospective employers. However, nothing materialized by the time I was posted from Karachi to Kakul in May 1971.

Meanwhile, the military regime had carried out mock negotiations with Mujibur Rahman. all the while preparing for military action in East Pakistan. This commenced in late March, and led to a bloody conflict that lasted for most of that year. The issue of atrocities in East Pakistan is a touchy one; most Pakistanis either deny them or would prefer to forget them, while in Bangladesh it is still an open wound. From my vantage point, it was clear that horrible things happened in that conflict, perpetrated by both sides against each other.

The wanton killing of civilians was initiated by the army when it launched the military operation in Dacca on the night of 25 March 1971; these were either people who resisted the military's seizure of key installations or Awami League cadres. In other places, the killings were mostly initiated by Bengali soldiers and paramilitaries, who killed not only their West Pakistani officers but also their families. These atrocities created a vicious cycle of uninhibited violence fed by an upsurge of great anger and hatred on both sides.

At the start of hostilities someone in Rawalpindi imposed censorship of all mail coming from East Pakistan. For some reason some of these letters landed on my desk. One such letter from a naval officer in Chittagong to his family in Karachi provided a graphic account of what was going on. Dated April 12, 1971, it makes for chilling reading, as these extracts show:

> "I am still alive and happy!.......Things out there in the (late) city have been pretty bad ('late' because it no longer exists).....Navy chaps got a chance to blast the goddamned Bingos after a long and miserable wait. When we could no

longer hold out in the port, we requested help from the army. And, oh boy, they came in with a bang that these Bingos will remember as long as they live. The air force was also very generous and spared a couple of thousand-pounders for the enemy. PNS Jahangir also carried out a bombardment.

I am commanding a platoon from PNS Dacca. No bragging, but we have done so much killing here that everyone calls us THBS (Trigger Happy Bastards Squad). I have so far killed about a dozen of these rotten Bingos.

The army has suffered heavy casualties so far, with many of their senior officers who were commanding Bengali units being brutally killed, along with their families, when the units mutinied. Many of their bodies have not been recovered yet; those that have been are in such terrible shape as cannot be described. A number of my dear friends have also been brutally murdered. The paper mill in Chandargaon had about 300 West Pakistanis working in it; every single one of them, along with their families, was killed and their bodies thrown into the river.

Everyone is saying we should kill as many of these Bengalis as possible, for what they have done to our people. The army is in a terrible mood these days. The navy has captured hundreds of prisoners, but the army does not believe in doing so. They do capture them, but will finish them off at the first opportunity. The entire Chittagong city is full of dead bodies, most of them rotting and stinking badly. The Karnafuly River is full of floating dead bodies. No one can roam about without a hanky tied across his face.

I believe this is the general situation all over East Pakistan. Don't believe a word of what Radio Pakistan says".

However understandable these primitive feelings of revenge may be in those circumstances, they can in no way justify or excuse the actions of officers, especially senior officers, of a disciplined military in permitting, much less encouraging, such atrocities. But such standards were hard to maintain with commanders such as AAK Niazi at the top. This was perhaps the biggest crime he committed, greater even than the cowardly incompetence he displayed as commander of the forces in East Pakistan, leading to their shameful surrender to the Indian Army.

3

Days of Shame

In May of 1971 I was promoted to the rank of Brigadier and posted as Deputy Commandant in the Pakistan Military Academy in Kakul. Apart from a couple of visits many years ago to witness passing out parades, this was my first stay in the PMA since the year I had spent there in 1948 as a cadet. Much remained unchanged, and it was an odd feeling being there again in this very different capacity. One of the new items was the Deputy Commandant's residence, a nice, comfortable house on a ridge overlooking the academy complex. This compensated a little for the fact that there was not much for me to do: I was supposed to oversee the academy's administration, but the settled routines that had developed over the years did not require much oversight.

A couple of months after my arrival we were told that President Yahya Khan was coming for a few days' stay in the PMA guesthouse, which was next to my residence. I had been told that it was built there on his specific direction for use on these occasional visits. A glass-fronted outer room had also been added at his direction, it was said, where he liked to sit up all night drinking while contemplating the twinkling lights of Abbottabad on the hills across the valley.

He duly arrived one evening in a convoy of cars. Sometime later I was informed by his Military Secretary (the pious, five-times praying Maj Gen Ishaq) that a lady would be arriving to visit the President later and her car should not be stopped at the Academy gate. I learnt later the visitor was one of his paramours from Rawalpindi. He left after a few days, and as the Commandant and I bade him farewell he said something about Pakistan being at war in a few months. (He also proved to me that the story I'd heard about his nightly vigils was correct when, out of the blue, he asked me how much I ran! It took me a few moments to realize that he was referring to my daily early morning run in the sports field below the ridge. Obviously, he had been watching me from his 'drinking room').

We were told by GHQ to make arrangements for running short courses to produce more officers for the army. The regular senior course, which still had several months to run, was also ordered to be wound up early, and the cadets passed out. The Commandant was away at some GHQ conference and asked me to take the salute at this passing out parade. I thus became the first graduate from the PMA to take the salute at one of its passing out parades[3], normally the preserve of top national and military leaders!

In early November I was told to immediately take over the artillery of 6 Armoured Division as its commander was medically incapacitated. Within a couple of days I joined my command and discovered that, though the division was in Gujranwala, my HQ was in the Narowal area, near the Indian border, where it was commanding an artillery group of seven artillery regiments (of different calibres from field to heavy)

---

[3] I recently discovered that among the faceless ranks that marched past the reviewing stand that day was Gentleman Cadet Ashfaq Parvez Kayani, who later became (and was until recently) the Army Chief.

The calm before the storm

Passing out parade at the Pakistan Military Academy
28 August 1971

On the reviewing stand

The family: (*L to R*: Yawar, Jamila, my mother, Afroze)

assembled from several formations to support an attack on the Dharam enclave, a piece of Indian territory on our side of the river Ravi.

The war started on the evening of 3rd December 1971 with the launching of our attack on the Dharam enclave (along with a few isolated attacks in other sectors, and air attacks on some Indian airfields) commencing hostilities on the West Pakistan border with India. Our attack succeeded with just the preliminary bombardment from our guns; the Indians withdrew and our infantry occupied the enclave without a fight. With no signs of any Indian riposte or even of any defences or troops on the other side, I urged the local division commander to exploit the success by pushing on into Indian territory. He said he had strict orders not to do so. The same thing happened in other sectors, too. It seemed the only purpose of starting the war in the West was to push foreign powers to intervene and impose a ceasefire in both East and West.

I rejoined the division with most of my units after a couple of days. In Gujranwala I discovered that, as well as commanding my own divisional artillery, I was to be the artillery commander for the newly formed Army Reserve North (consisting of our division and 17 Division) and also the *de facto* artillery adviser to HQ 1 Corps (which was in charge of the divisions defending the Silakot – Shakargarh - Narowal sector), since its artillery commander had parked himself 40 miles away with his HQ and guns in order to carry out counter-bombardment (though I suspect it was mainly to get away from the corps commander). The Army Reserve commander was based at our division HQ with a couple of staff officers (it was generally believed that he was really there to keep an eye on our division commander, who was a Bengali officer). My multiple roles not only allowed me to observe the progress of the war throughout this area, but also gave me the opportunity to try and influence it.

This I did by trying to push the Army Reserve commander, Maj Gen Bashir Khan, into planning for possible operations that the Army Reserve might need to carry out, so that I could plan the artillery support for them. However, he seemed at a total loss and would not commit himself to anything. Discussing this one day with Abad, the GSO-1 (the senior operations staff officer) and Aleem Afridi (Gen Bashir's chief of staff, and an old friend) in the division HQ operations dugout I sketched out on the map a plan for the Army Reserve to deal with an enemy push along their most likely axis of advance. Just then Gens Bashir and Karim walked in, and Aleem showed them the plan on the map. They both expressed agreement with it, but when Aleem later on asked Gen Bashir if he could issue an Operations Instruction on that basis, Bashir refused to let him; obviously he didn't want to commit to anything in writing.

However, for me that verbal agreement, however tentative, was enough to begin my own planning to assemble as large a force of artillery as possible to support this or a similar operation. I spent the next day visiting various artillery commanders in our sector to ask them what units they would be able to provide to support an Army Reserve operation. I found that I could count on 14 artillery regiments for this purpose, and accordingly prepared and issued an operation order to the various formations to carry out the planning and preparation necessary to assemble this mass of artillery. As a result of these preparations all this firepower could have become available with a few hours' notice from the night of 11 December onwards.

Meanwhile, the battle was going badly for us along the border. The Indians had now deployed a substantial force there and slowly pushed into our territory, with more and more of it being lost every day. The soldiers fought well, but their commanders were tragically inept. The way the battle was being mismanaged was graphically brought home to me when I happened to visit HQ 8 Division on 8 December when a counterattack was to be launched by 8 Armoured Brigade to deal with an Indian penetration. I wrote the following account soon after the war ended while it was all still fresh in my mind:

> "I found the division commander, Maj Gen Abdul Ali Malik, sitting in the operations bunker with Brig Muhammad Ahmad, commander 8 Armd Brigade. The general atmosphere was gloomy and tense; the staff, all of whom seemed to be present, appeared to have been up all night. Some were unshaven, all were weary, haggard, red-eyed. An attack was to go in that afternoon and they were just awaiting the release of 8 Armoured Brigade by Commander 1 Corps. At 1330 the latter called Gen Malik and released the brigade.
>
> I had till then presumed that the attack, due to begin at 1530, was all planned and prepared, but after Brig Ahmad hurriedly left, all hell broke loose! One staff officer began drafting the request for an air strike at 1430, another for a CAP (combat air patrol) at 1500; the Col Staff summoned the Commander Signals to arrange communications with the armoured brigade but when he arrived he said that it couldn't be done on an hour's notice.
>
> An Army Aviation pilot came in asking for the radio frequencies of the two regiments of the brigade taking part in the attack but no one knew, so he sat on the floor and rang up the brigade major, who also didn't know. The GSO-1 read out his air support request to the GOC and got a 'rocket' for his pains. The commander artillery was away

making the fire support plan, God knows with whom since the brigade commander had been at the Div HQ. It was pitiful!

I quietly stole away to visit one of my regiments deployed in support of the brigade holding the Shakargarh sector.

On my way back I stopped at HQ 8 Div at about 1700 to find out the results of the attack. They were all sitting there looking dazed, and didn't know what was happening. Later on I learnt the fate of the attack. Two armoured regiments were to be launched in a big pincer. They had to move 40–50 miles to their assembly areas, and the attack started late. One regiment got held up 1000 yards from its start line; the other advanced a few thousand yards without meeting any opposition. Then it began to get dark and everyone pulled back. Apparently, the minefield gap was closed too hastily and some troops were left stranded on the other side. We lost two tanks and some men, and gained nothing. Four enemy tanks were claimed but I have my doubts".

This attack was typical of the way operations were conducted throughout the war, at least in this sector. Much bloodier fiascos occurred in the closing days of the war with 13 Lancers and 35 FF. This latter battalion was moved by rail from Hyderabad to our area. They reached their assigned brigade's sector after spending almost 24 hours on the train. The same night they were launched in an attack on an enemy penetration – without any planning or preparation, liaison with frontline units, reconnaissance or arranging fire support. They ran into a minefield, neighbouring units opened fire on them and called down artillery upon them. Many men were killed, including their CO. It was a massacre. The brigade commander responsible for this was Brig Sher Ali Baz, who had spent almost his entire service in the Military Police, and acquired a dubious reputation for torturing people in Baluchistan (Nawab Akbar Bugti referred to him once as "a specialist in the art of torture").

The overall strategy being followed was no better. On 12 Dec the army Chief of Staff, Gen Hamid Khan, came to HQ 1 Corps to confer with the local commanders. I had accompanied Gen Bashir, the Army Reserve commander, to Corps HQ, and was in the room adjoining the one where the generals were meeting when I was suddenly called in. Gen Bashir introduced me by saying to Gen Hamid: *Brig Ali has some good ideas on the subject*. He then asked me to brief them. Unsure of what he wanted me to talk about, I began to explain the plan I had sold to Gen Bashir for his potential operation. Gen Hamid obviously wasn't interested in this and instead asked what I thought should be done in the present situation. This was an unexpected question, but I immediately replied that the artillery must be used in a concentrated fashion. I explained that 14 regiments from within the sector could be assembled but if he gave us some Corps artillery units from neighbouring sectors that were not engaged, we could bring about 20 regiments of artillery to bear upon the enemy which would stop any advance. He said he couldn't do that, but directed the Corps commander to concentrate his own artillery. In the event, nothing happened.

The meeting soon broke up rather hurriedly as it was time for the pre-arranged air cover for Gen Hamid's helicopter flight back to Pindi. As he was leaving Gen Hamid took me aside and said, *If you have any more good ideas, let me know; ring up my Military Secretary*. This was how our 'higher direction of the war' was being conducted.

Another example of the manner in which this 'higher direction' was functioning can be gauged from another incident.

One day the Chief of the General Staff rang up my division commander and ordered him to move some AA guns to Gujrat (a small town of no military significance well back in the interior). Since these guns were under my command, the order was passed on to me. All these AA guns were deployed in the battle area where the Indian Air Force was very active, and were already too few to meet the need there. Accordingly, I told Gen Karim that I would arrange for some other guns in the rear areas to be redeployed to Gujrat. He agreed, but next day received a dressing down from the CGS at the delay in carrying out his order. I had no choice but to move my guns. The only reason I could think of for this ridiculous order was that this town happened to be where Gen Yahya's 'madam' (the notorious 'General' Rani) lived.

I became more and more frustrated with the generals' unwillingness to fight, and their refusal to let anyone else do so either. Every effort of mine to have my divisional artillery join the battle (arguing that artillery was never kept in reserve) was vetoed by GHQ. My frustration boiled over one day in a conference at HQ 1 Corps. Its commander, Lt Gen Irshad Ahmed Khan, was not only utterly incompetent but a coward as well. He spent more time assuring his personal safety than he did commanding his formation. (The Hamood-ur-Rahman Commission after the war rightly recommended that he be court-martialled for his disgraceful performance in surrendering nearly 500 villages to the enemy without a fight).

During this meeting (on 14 Dec) I pointed out to him that large tracts of territory were being lost to the enemy without any resistance being put up, and asked him when he proposed to recover all this territory considering he had four armoured regiments available. He replied that he actually had six regiments, but wasn't intending to recover any territory, saying:

Under the GHQ plan I can give up 20 times more territory. This battle is going according to plan! At this I lost my cool, and said in a rather rough tone: *If you're not going to use these regiments, why don't you give them to us so that we can try to recover the lost territory*? After a few moments of shocked silence, he burst out: *Don't forget that after the war you will come back under my command, and I will write your ACR* (annual confidential report).

Finally, on 15 Dec I decided to place on record what I had been urging upon the generals all this time. I wrote a formal letter to Gen Bashir and the other generals pointing out the great artillery firepower that was available but not being used (a copy of this letter is attached to the brief at Appendix 2, page 57). This was also the day on which Radio Pakistan admitted for the first time the dire situation in East Pakistan, which was also gradually becoming known in the army. Early that morning, as I was shaving, my Signals officer, an ex-SSG (commando) major, burst into my caravan weeping loudly, crying out that his friends were dying in Dacca and we were sitting here doing nothing; he wanted to go there and die with them. It took a while to calm him down.

On 16 Dec at around midday we became aware of a crisis in 8 Div area — the enemy had breached the last minefield NE of Zafarwal and attempts to eliminate his bridgehead had failed. I began pressing Gen Bashir to let me move my guns forward as they were likely to be needed, but he told me at about 1530 that GHQ had refused to allow any moves forward. However, three hours later, GHQ ordered the whole of the armoured division to move immediately into the battle area (even though they had repeatedly assured us we would get 48 hours' notice for such a move)! As I later wrote in my submission to the Hamood-ur-

Rahman Commission: *Only a military man can savour the full flavour of folly and incompetence contained in this order.*

I spent the next morning at HQ 8 Div sorting out with their artillery commander where he wanted me to deploy the Army Reserve artillery, and issuing the necessary orders. The Corps commander and several other generals were at the HQ in a meeting about the serious situation on the front. They were apparently quite panicked; Lt Gen Irshad spoke to the army Chief of Staff for 20 minutes and then for 10 minutes to the President. I can imagine what he told them.

At about 1 p.m. as I was talking to Kamal Rabbani (8 Div artillery commander) in the operations bunker an officer burst in with the news that we had accepted an Indian ceasefire offer! We refused to believe it until a couple of others confirmed that they had also heard it over the radio. It was a stunning shock. The first reaction was rage, a consuming, devouring rage. Then came contempt, contempt for these generals who did not know how to fight, and would not let others who did do so, and who had now lost their nerve even though the bulk of the army was still intact, indeed had hardly done any fighting at all. Coming a day after the shameful surrender in East Pakistan which made about 90,000 of our soldiers Indian prisoners of war, this was a crushing blow.

Five generals were sitting in the adjoining dugout eating lunch. I turned to an officer and suggested he send in some roast chickens, as their appetites would now have revived and they might feel like celebrating. Just then the big, bold headline on that day's newspaper caught my eye; *The War Still Goes On*, a quote from Yahya's radio broadcast the previous night on the Dacca surrender. *We might at least have been spared the final indignity of that speech*, I remarked to Kamal.

The drive back to my division HQ was a dark passage. In my final statement before the Attock court-martial (in Sept 1973) I recounted some of what I felt during the war and after:

> "No one who has not seen the refugees can comprehend the full disgrace of the 1971 war. In the battle area there was an unending stream of uprooted, destitute, bewildered humanity, clogging every road, every path, every track. They didn't complain, they didn't beg, they didn't even look at you, yet the reproach in their mere presence was a crushing burden. For 24 years they had nurtured and fattened an army with their sweat and blood. Our salaries, our bungalows and cars, our polished jeeps and guns and tanks, all had come from the toil and sacrifice of these poor people. And now, in their hour of need, we were sitting around in comfortable dugouts, hiding under trees, while the enemy took over their homes, captured their women and children, and cast them out to wander the countryside, lost and helpless. Anyone who had a conscience could not but feel the terrible anguish of his personal responsibility............
>
> December 17th seemed to me like the end of the world. I felt that the country, the people and the armed forces had all been betrayed. What made it worse was that, though I felt personally betrayed, I realised that I, too, bore some of the responsibility for what had happened. Every senior officer was responsible for the shame and disgrace that had fallen upon us".

I decided to accept my share of this responsibility. That evening, shortly after the ceasefire came into effect, I drafted and signed my letter of resignation from the army in which I had served for 24 years and sent it to the appropriate staff officer at the HQ.

4

The Day of Redemption

When I sent in my resignation on the evening of the 17th December I took it for granted the Yahya regime would realise that, after the horrible debacle they had caused in both East and West Pakistan, they could no longer continue in power and would quit. I was astounded the next morning when I learnt that they had no such intention; in fact, Yahya announced that he was going to promulgate a new constitution! Meanwhile, large-scale public demonstrations against the regime had broken out all across the country with people demanding it should go.

That morning (the 18th Dec) our division commander called a meeting of his three Brigadiers and the Colonel Staff. Maj Gen Karim was a Bengali officer from what had been East Pakistan, and he wasn't sure where he stood after the recent events. He explained this dilemma and asked if we still had confidence in him, failing which he would ask to be reassigned. As the senior officer present I replied first; I told him I and the men under my command had full confidence in him but we had none in anyone above him. The others in turn expressed general agreement with my views.

This led to a general discussion about how things stood in the aftermath of the war. I expressed serious concern that the regime might use the army to put down the ongoing public agitation against it. With the army in its current frame of mind its discipline might break down under such a severe strain. I said that there was no way out except for the regime to go; the others felt the same way. We urged Gen Karim to convey our views to GHQ. He seemed to agree with us and said he would think over the best way of conveying this message.

The morning turned to afternoon and still Gen Karim took no action. Finally, in late afternoon Col Javed Iqbal (the Col Staff) and I drafted a signal[4] for Gen Karim to send to Lt Gen Gul Hasan, the Chief of the General Staff. Javed took it to Gen Karim in his caravan; the latter was quite taken aback, remarking that it amounted to sedition; he asked Javed to leave it with him and he would think it over. But the day passed without him doing anything.

Next morning, the 19th, I discussed the impasse with Javed Iqbal and decided to force Gen Karim's hand. I drafted a formal

---

[4] This signal (which was never sent) is shown in Appendix 5 of Brig AR Siddiqui's *East Pakistan – The Endgame* (Karachi: Oxford University Press, 2004), and reads:

All my formation commanders categorically state that they and their officers and men have lost complete faith in the present government and army leadership. They state that they and their troops will not take up arms against the people to protect the present regime. In case of resumption of hostilities will not fight under present military direction. To preserve integrity and cohesion of country and army absolutely essential immediate changes be effected in regime and military leadership. Government must be handed over to leaders acceptable to the people. Events causing acute dissatisfaction among officers and men and situation fast deteriorating.

(Appendix 7 of the book contains the transcript of a letter that Gen Karim wrote to Gen Gul Hassan, dated February 2, 1972, dealing with these events; in it he tries to protect me and some of the other officers involved by minimizing our actions. The book itself does not refer, apart from a passing mention, to the events of 19 Dec. The letter and this signal were obviously provided to the author by Gen Karim).

official letter from me to Gen Karim repeating what was contained in the signal of the day before, and requesting that my views and those of my command be conveyed to "the quarters that can take action on them". Javed had it typed up and I signed it; we then went to see Gen Karim. He started off by saying he considered it best to convey our message through a personal letter to Gen Gul Hasan, and read out what he had written; the hand-written letter concluded with a request for 'talking points' to be issued for commanders to use! I told him this was quite insufficient, and suggested that, if he booked off sick, I would take over acting command as the senior officer and would send the signal we had proposed earlier. He did not agree to this.

I then took out the letter I had prepared that morning and handed it over to him. He read it and went rather pale. He asked who had typed it; Javed said his PA had done so. He was silent for a while, then his manner turned formal and official and he said very coldly: *Very well, I will let you know my decision. That will be all.* We came out of his caravan and stood nearby, not sure what to do next. I had wanted, through my letter, to push Gen Karim into acting, but it seemed that what he was going to do was act against us. All that we were trying to achieve was in danger of being lost, but I couldn't think of what to do next.

Just then Gen Karim came out of his caravan and, pulling on his gloves, walked towards his waiting jeep. I realised that once he departed it would all be over: everything lost for no gain whatsoever. With hardly a moment's hesitation I went up to Gen Karim, saluted, and said: *I'm sorry, sir, but I'm taking over command of the division*! He just looked at me, speechless. I walked away with my mind in overdrive, trying to figure out what I should do next; I had never even thought about this eventuality. Just then I saw the division signals battalion commander, Lt Col M Khurshid, and, calling him over, told him to immediately cut all communications out of the HQ. Without any question, he ran off towards the switchboard tent to do this. When he came back I told him I had taken over the division and wanted him to have three young officers ready with weapons nearby in case I needed them.

Shortly after, Gen Bashir, the Army Reserve commander, arrived and went up to Gen Karim, who was still standing outside his caravan. Presumably Karim told him what had happened. Bashir then came towards me, shouting: *What do you think you're doing? You are under arrest! Go and sit in my jeep and I will take you to corps HQ.* I said to him: *No, sir*, and called out to Khurshid to get the officers. When Bashir saw the three young officers running towards us with their weapons, he suddenly changed his demeanour, saying: *All right, all right, let's talk about this.* Gen Karim joined us and also sought to defuse the situation, telling someone to put some chairs out in the sun.

In a little while a small group had gathered in this circle of chairs in the morning winter sun: Gens Bashir and Karim, Maj Gen RD Shamim (commanding the other division in the Army Reserve, who had just happened to arrive then), Brig Iqbal Mehdi Shah (commanding an armoured brigade in our division, whom I had asked to come over), Col Aleem Afridi (Bashir's chief of staff), Javed Iqbal, Khurshid, Lt Col Bashir Butt (a retired officer recalled during the war and attached to our HQ), and myself. The lively discussion was mostly led by the younger officers, while I ensured that it moved towards a practical approach for achieving the goals we all seemed to agree on, namely, that the regime had to go, together with all the cowardly and incompetent generals who had led us into this defeat.

The consensus arrived at was that we would adopt these goals as our stand, but take no overt action to achieve them. By conveying this stand to the regime, we would compel it to seek the support of the rest of the army; we were confident this would not be forthcoming, thus forcing the regime to quit. The other major issue that came up was to whom the regime should hand over power. Various views were expressed but I was very firm that we should express no preference and that the only legitimate successor was the party that had won the last elections in West Pakistan. I knew this meant ZA Bhutto and his People's Party. I was no fan of either, being fully aware of their dark side, but felt that the military had no choice but to support them as they had been elected by the people.

The next question was how to convey our stand to the regime. The generals suggested we ask Lt Gen Gul Hasan, Chief of the General Staff, to come to our HQ to be briefed. Gen Bashir and Javed Iqbal spoke to him on the phone but he said he could not leave GHQ, and asked for someone to come and see him. We decided that Aleem Afridi and Javed Iqbal should go. Gen Karim gave them the letter to Gul Hasan he had drafted the day before (the one we had found quite inadequate) and Gen Shamim also sent a letter expressing his feelings. I told the two Cols that the real message they should convey was the stand we had adopted, and to add that if an announcement of its acceptance was not made by 8 p.m. we could not guarantee control of the situation and what might then happen. They left by helicopter around noon.

When they reached Gul Hasan's office (as I learnt later) they found him in a very despondent mood, clearing out his desk. He said he was being blamed for the defeat, and had decided to resign. However, when he heard their account he cheered up and rang his friend the Air Force chief, Air Marshal Rahim Khan, to come over at once. When he arrived, Aleem and Javed repeated their story and the demand they had come with. Gul Hasan and

Rahim Khan then went to see Gen Hamid, the army Chief of Staff, leaving the two Cols cooling their heels in the CGS offices. It was only later that night that they learnt that, when our demand was conveyed to Gen Hamid, he arranged for Gul and Rahim to see the President at 7 p.m.

Back in the Div HQ I was expecting a call from Aleem and Javed as to how their meeting with Gul Hasan had gone but heard nothing, leading me to fear the worst. Matters were not helped when, sometime later, Gen Karim sent Lt Col Bashir Butt to warn me that Gen Bashir (who had planted himself in Karim's caravan) had spoken to Gen Hamid in GHQ, and I should be careful. The Signals CO, Lt Col Khurshid, was a very fine fellow but of a rather nervous disposition; during the afternoon and evening he came to me with several alarming reports. I began to fear that Gen Hamid might mount an operation against our HQ to nip our action in the bud. Accordingly, in late afternoon I asked Brig Iqbal Mehdi Shah to send an infantry company to the HQ, and they arrived within a couple of hours.

Sitting alone in my caravan with no news all these hours, I grew increasingly anxious. As darkness fell, the pace of worrying signs picked up. I had given orders that all calls for the two generals be routed to me instead. Several came, including one from Lt Gen Tikka Khan, but when I told the operator that the generals were not available but he could speak to me, he rang off. At one point Khurshid came running, saying a call from GHQ had got through to Gen Bashir and he had ordered the helipad to be lighted up; I told him to have the lights extinguished, fearing that SSG commandos might be brought in by helicopters. (Later on I learnt that Gen Mitha, a regime stalwart and former SSG commander, had in fact tried to get hold of a company of SSG, but was unsuccessful).

Around 7.30 p.m. I got a call from the telephone exchange

supervisor saying the President wanted to talk to Gen Bashir. I told him to hold the call and walked across to Gen Karim's caravan nearby, where the two generals were sitting. I urged Gen Bashir to take the call and tell the President about the stand we had taken. He literally begged me to spare him. By the time I asked the operator to let me talk to the President instead, he had rung off. Apparently, this was after Gul Hasan and Rahim Khan had conveyed our demand to Yahya Khan. Hamid had already told the President that he had been unsuccessful in getting any other formation or troops to act against us. The failure to get Bashir on the phone probably ended their options. Yahya told Gul Hasan to have the broadcast made about his intention to hand over power to the elected politicians. Shortly before our deadline of 8 p.m. the news came out over the airwaves.

The relief was short-lived. The cabal around Yahya had not yet given up. Shortly after, Gen Mitha spoke to Gen Karim and tried to persuade him that Gen Hamid should take over from Yahya; Karim told him to talk to me. He tried to sell this pitch to me but I told him, firmly but respectfully, that it was not possible. A little later we got a message that the DG ISI (Director General, Inter-Services Intelligence Directorate) and the DMI (Director of Military Intelligence) were coming to visit our HQ and we should send a vehicle with a guide to Gujranwala airfield for them. I had Khurshid send one of his officers, and for the latter to report who he was bringing with him on his return. This created another scare later on when Khurshid came to me in a panic saying his officer had been arrested at the airfield. Later on this was cleared up as a misunderstanding.

From the presidency Gul Hasan and Rahim went to the former's residence, and celebrated their success with a few toasts. Gul then turned up at his office and, in great good humour, briefed Javed and Aleem on what had transpired; at one point he laughed so hard he fell off his chair! Then he rang our HQ, spoke to Gens Bashir and Karim and they congratulated each other (I was nearby). When Gul was told the DMI was coming to our HQ he was very surprised and angry, and cursed him royally; apparently Iqbal had been sent by Hamid and went off without telling Gul, his boss. Gul Hasan also said he would come to our HQ the next morning to "thank the officers and men".

Brig Jilani, the DG ISI, and Brig Iqbal, the DMI, arrived at our HQ around 10 p.m. in quite a nervous state, unsure of the reception they would receive. They were taken to the operations caravan where a group of officers soon assembled, comprising Gens Bashir, Karim and Shamim, Brig Iqbal Mehdi Shah, and me. There was a general discussion of the war and the situation following it, in particular how unhappy the army was with what had happened. Gen Karim then said that his division and Shamim's had adopted the stand that the military regime had to go and hand over to a civilian government; Gen Shamim confirmed this. The visitors didn't comment on this, instead Jilani just rambled on about how he felt the same way as we did. Karim had to interject: *Yes, but we have adopted a stand*! I kept studiously quiet, and did not participate at all; I wanted to assess what these visitors were up to, but it appeared they were there just to check things out. Towards the end of the discussion, Brig Zia-ul-Haq suddenly appeared and joined us; I do not know what Zia was doing there. He was posted to another formation about 100 miles away but had suddenly appeared at our HQ on the 18th, and remained in our sector till the 20th, moving around the area. He was a great friend and protégé of Gen Gul Hassan, and I assume he was scouting around on his behalf.

After this rather aimless discussion petered out, Brig Iqbal, the DMI, got hold of Iqbal Mehdi Shah and took him outside. He spent more than half an hour talking to him very earnestly, which caused me some anxiety. After the two visitors left, Iqbal Mehdi Shah rather sheepishly asked me whether he could send his infantry company at the Div HQ back to their unit right away; I realised that this was what the DMI had browbeaten him about. I replied that it would be more appropriate if they went back next morning and, after another half-hearted attempt, he did not press the issue.

The DMI's determined effort to have the company guarding our HQ removed caused my earlier fears to return. It seemed likely they had prepared some kind of operation against us for that night. I went to bed at about 1 a.m. reconciled to the possibility that I would not see the next morning, and spent some time silently saying my good-byes before dropping off to sleep. (My fears were well-founded, as I later learnt. From our HQ the DMI went to an infantry battalion of 17 Div camped nearby whose commander was a friend of his. He tried to prevail upon the CO to use his battalion to take over our Div HQ, but the colonel expressed his inability to do so. Apparently, Iqbal then decided to switch sides, and next morning reported to Gul Hasan instead of the army Chief of Staff, Gen Hamid, who had sent him. I got a frantic call from Hamid's military secretary in the morning wanting to know the whereabouts of the DMI).

That morning, the 20th, Gen Karim made preparations for the promised visit by Gen Gul Hasan. He had the helipad prepared, and told me to call up some anti-aircraft guns and deploy them around the HQ for its protection. However, soon after, he got a call from Gul Hasan saying he couldn't leave Rawalpindi that day. We learnt later he had to attend the meeting of garrison officers that Gen Hamid had called (at which Hamid was hooted off the stage by many officers), and also because ZA Bhutto was arriving from Rome (where he had stopped on his way back from the United Nations, apparently to see which way the wind was blowing back home).

Hamid's debacle at the officers' gathering was the final nail in the coffin of the Yahya clique and their hopes of hanging on to power by having Hamid replace Yahya. When Bhutto landed, GM Khar met him on the plane (possibiy with Air Marshal Rahim) and briefed him. Emboldened, he brushed aside Yahya's half-hearted attempt to win some concessions for his henchmen and took over as President. That evening he made a broadcast to the country in which he announced that he had retired all the generals of the Yahya gang "in accord with the sentiments of the Armed Forces and the younger officers". He also made Lt. Gen. Gul Hassan the Army chief, and confirmed Rahim Khan as the Air Force chief (they had probably told him they were the ones who had engineered Yahya Khan's ouster).

After it was over I went to Gen Karim and apologised for having had to take over the division from him. I told him he already had my resignation; I would understand if he were to forward it to GHQ. No, no, he replied, I will tear it up.

Looking back on the events of the 19th, I am still amazed at how it all happened. Nothing was planned, I hadn't even thought about any such course of action. Once I took that momentous first step of telling Gen Karim I was taking over the command from him, it was as if I was borne upon a rushing torrent that carried me forward. All I had to do was make decisions as each problem arose and, amazingly, they all worked, every single one of them. Many problems arose that I never even had to deal with as they somehow resolved themselves (as I was to discover afterwards). Later on, recounting these events, I told people it seemed as if that day I was in the hands of God!

One of the most remarkable features of this event was the support and assistance I was given by the other officers who played key roles in it. Apart from Aleem Afridi I had not known them previously; where I did, the relationship had not been good. The key support was provided by Brig Iqbal Mehdi Shah, as he was the only one of us who commanded combat troops (I did, too, but my artillery regiments were not much use in this type of situation). We had served together in a station as Lt Cols and had not got along in our few interactions. Yet, when the moment came, he answered my call unhesitatingly; testament both to his essential nobility and the awfulness of the predicament in which we all found ourselves.

Another key person was Lt Col Khurshid, the divisional signals commander, who controlled communications from and to our HQ. I first met him a couple of days earlier when Aleem Afridi introduced us; he came across as a very sincere and upright officer, and was vocally upset at our defeat and angry at those who had caused it. Similarly, I had met Javed Iqbal for the first time during the war, and then only briefly. I was very gratified when, unasked, he gave me his invaluable support and assistance after seeing the position I took in Gen Karim's Dec 18th meeting. Without Javed and Khurshid I could not have assumed control over the Div HQ.

The true significance of the events of 19th and 20th December 1971 is that it was the Pakistan Army itself that had rid the country of this appalling regime that had ruled in its name. The overt action was taken by a small group of officers, but it depended for its success on the tacit support of the rest of the army. If even a small element of the military had acted to preserve the regime, our move could well have failed since I was determined there would be no fighting within the army. Even so, this does not diminish the great credit due to these officers who followed my lead. They risked everything (their careers, their liberty, their families, even their lives) to answer the call of this critical moment in their nation's destiny. (Apart from those named above there were also many others who knew what we were doing and tacitly supported us).

5

The Days After

Within a week of taking over as President, ZA Bhutto established a commission of enquiry under the Chief Justice of Pakistan, Justice Hamood-ur-Rahman. Its mandate was to enquire into the 1971 war debacle, affix responsibility and make recommendations. The Commission published a notice in the press inviting submissions from anyone who had any information relevant to its terms of reference. I was thrilled at this opportunity and prepared a submission which I sent in on January 2, 1972 (reproduced at Appendix 2, page 50). Later, I appeared before the Commission twice to testify about the war and the events surrounding Yahya Khan's ouster.

Meanwhile, I had been awaiting the fulfilment of the second part of our demand of Dec 19th (i.e, getting rid of the incompetent and cowardly army commanders). However, it soon became apparent that the new army chief, Gen Gul Hasan, was instead going to settle some personal scores while not touching those responsible for the war debacle. He ordered the retirement of three generals who had no connection to the Yahya regime and who had not been involved in the war failures. In fact, they were all considered very good officers, professionally proficient and with clean records. I was particularly upset because two of them (Maj Gens Ihsanul Haq and Shaukat Riza) were friends, especially the former whom I had known for a long time.

After stewing over this for a few days, I decided to bring it to the notice of the new regime. The simplest way seemed to be to go and see Ghulam Mustafa Khar, newly appointed Governor of the Punjab, and a right-hand man of President Bhutto. Accordingly, one morning I drove from the field area to Government House in Lahore, a palatial building with huge grounds built in the days of the British *Raj.*

When I told the Governor's secretary I wanted to see his boss, he told me he didn't think it would be possible as Mr Bhutto was staying there and the Governor was busy with him. I pressed him to at least tell Mr Khar that I wanted to see him on an important matter. His reluctance was overcome by the urging of another person sitting there, whom I later discovered to be Hanif Ramay, a senior PPP leader. After a little while the secretary's intercom buzzed and he ushered me into Khar's office, a huge room, at the far end of which sat a small figure behind a big desk.

After introducing myself I narrated to Khar a brief account of the events of Dec 19th, and then told him that Gul Hasan appeared to be misleading the government and it would be blamed for the failure to clean out the corrupt and incompetent generals still serving. He listened carefuliy and then asked me to wait in the secretary's office. After a short interval his military secretary came and asked me to accompany him. We stopped at his office where he took the pistol I was wearing and locked it in his safe, then led me to another small room upstairs and asked me to wait there.

After a few minutes Bhutto came in accompanied by Khar. Bhutto looked more like a film star than a politician. I stood and saluted; he asked me to sit down, and said, "Normally I don't get involved in these things, but my friend here thinks I should listen to you". I repeated what I had previously told Khar. When I told him about the unjustified recent retirements he turned to Khar and said (obviously for my benefit), "I thought there was something fishy about that case". When I finished he said he would consider what I had told him and, as he left, said to Khar, "Make sure that Military Secretary doesn't see him". (He meant the MS he had inherited from Yahya Khan. This remark was again obviously for my benefit; Bhutto just couldn't help being 'clever', even when there was no need for it). Khar asked me to wait downstairs.

When I got to the secretary's office, he was already on the phone trying to get someone in the President's House (in Rawalpindi) on the line for Khar. I gathered this was a retired army officer who was their party 'expert' on the military; obviously they wanted him to find out who I was. Before long Khar called me into his office again and said they would act on what I had told them; he also asked me to remain in touch. I then returned to the Div HQ.

Gul Hasan wasted no time. That evening Gen Karim summoned me and said he had received orders for my immediate transfer to the School of Artillery (as its Commandant). I was to leave next day without visiting my command to say farewell, or handing over to my relief who would be arriving later. I was not too surprised; I had known there was only an off chance that Bhutto would act appropriately on my information. Next day, after a few goodbyes on the phone, I left for Nowshera. A few days later I called up Khar and told him what had happened. He replied: don't worry, everything will be all right, stay in touch!

After a week or so I was informed that a court of inquiry had been set up by GHQ (under a Maj Gen) to investigate the events of 19-20 Dec and the conduct of certain officers, including me. The others were Iqbal Mehdi Shah, Javed Iqbal, Aleem Afridi,

Khurshid, and Gen RD Shamim. At the same time Aleem Afridi and Javed Iqbal received messages from Gul Hasan that we shouldn't worry as this was just to clear his own position and nothing would happen to us. I again called Governor Khar and brought him up to date. His response was the same: don't worry, all will be well, stay in touch. That was the last time I spoke to him. (As it turned out, Gul Hasan was the one who should have been worried; a couple of months later he and Air Marshal Rahim were unceremoniously ousted from their positions by Bhutto as they proved insufficiently pliable. Gen Tikka Khan replaced Gul Hasan).

The court of inquiry lasted for several months, during which Jamila and I would periodically travel to Rawalpindi where it was being held. All the other officers would be there as well, and we enjoyed these reunions (we stayed with Aleem and Shamim Afridi in Pindi; they were old friends from our earlier Nowshera days). The inquiry was also a lot of fun, especially being able to cross-examine witnesses such as Gen Bashir and having them make fools of themselves. Not that it made any difference to the outcome.

One evening in August I was sitting at home with my family watching a Bhutto press conference on TV when I was called to the phone. It was Gen Tikka Khan. I knew Tikka, in fact I had served as a staff officer under him many years ago. He said he was sorry to give me some bad news: I was being retired, and the news would be announced soon! I called Jamila out from the family room and told her; we returned in time to see Bhutto saying that his adviser, Rafi Raza, had an important announcement to make. All of us were being retired, Raza said, because we had engaged in a conspiracy to (of all things) *prevent* the handover of power to a civilian government in Dec 1971!

I did not much mind my career ending in this fashion. I had joined the army with the secret hope that one day I could make a significant contribution to the country. What I hadn't allowed for was that, meanwhile, the country would go to the dogs, and that by the time I got anywhere worthwhile it might be unsalvageable. That is why, despondent both at what was happening and my own impotence, I had decided the previous year to give up this hope and resign. The war intervened, and provided a last opportunity to do something that could perhaps make a difference. While we had stopped the rot under Yahya from proceeding further, the end result was to give Bhutto his turn to preside over the downward spiral.

While I didn't mourn the loss of my own career, I felt very sad at the ending of the potentially bright careers of the other officers who had acted so bravely and selflessly — Muhammad Khurshid, Javed Iqbal, Aleem Afridi and Iqbal Mehdi Shah were all retired. Gen Karim was not retired with us because, as an East Pakistani officer, his career was finished anyway. Gen Shamim also got the axe, even though he had backed off after that morning session on Dec 19 when he found out later on that his senior officers refused to go along. The commander of the company that Iqbal Mehdi Shah had sent to the Div HQ at my request, Major Afzal, and his battalion commander, Lt Col Qadir Saeed, both very good officers, also had their careers blighted merely because they had followed orders.

We packed up in a few days and left Nowshera with a minimum of ceremony, going to my mother's house in Lahore. It was a strange feeling to be at a loose end with nothing to do, and no future to plan or prepare for. After a few days Aleem and Shamim Afridi arrived in Lahore; he knew M. Anwar, one of the top lawyers there, and wanted us to file a suit against the government for wrongful termination of our careers. This was a welcome antidote to the initial inaction, and I got busy with him in preparing the case.

It was during this time that Aleem told me about an Air Force officer, Sq Ldr Ghaus, who had contacted him and told him there was a group of officers in the PAF who were very upset that Bhutto had not cleaned out the rotten layer at the top of the military and were talking of overthrowing his government to achieve this. About the same time Major Farouk Adam Khan (related by marriage to Shamim) also told Aleem that he and some friends in the army were also very upset about this and were thinking along the same lines. Deeply mortified at the defeat caused by these cowardly and incompetent senior officers, these young officers wanted a reform of the military, starting with their removal. But Bhutto displayed no interest in taking any such step; it suited him to have these tarnished generals remain at the top of the military, beholden to him for their survival.

The question of making a living also demanded my attention. I explored various options but, finally, Jamila's brother Ijaz insisted I come to Karachi and work with him. He was running a real estate development company but wanted me to start a new export business for him. With nothing better available, I agreed. Accordingly, after a couple of months in Lahore we moved to Karachi and settled down to begin a new life.

My attempts to start an export operation for my brother-in-law were a very frustrating experience. I researched available items suitable for export and then wrote letters to businesses all over the world who might want to import these goods, but no one seemed interested. The few parties importing things from Pakistan already had established suppliers and did not want to talk to anyone else; others who might have been interested had probably already been burnt by the many swindlers operating in this field. After months of fruitless effort, I realised I could no longer take advantage of Ijaz's generosity, and would have to find some other employment.

My personal frustration was compounded by political developments in Pakistan. Bhutto was consolidating his power and establishing a dictatorship: repression increased, as did the subversion and undermining of the few remaining national institutions (the judiciary, the civil service and the armed forces). My part in enabling this man to ascend to power began to gnaw at me. I tried consoling myself with the thought that he would have got there anyway, and probably not in such a bloodless fashion. After all, in his quest for power he had not hesitated in contributing significantly to the breakup of the country with all its attendant horrors. He would have used any means, including his popular support, to wrest power, and that would likely have involved much bloodshed.

What a glorious opportunity Bhutto had when he became President. The people of Pakistan were shaken down to the roots of their national psyche. The country had splintered, but much worse was that the very basis of their nationhood, their justification for being one people, long under strain, had finally shattered. Their lives, devoted mainly to selfish, individual pursuits, suddenly stood starkly revealed in their pettiness. Shorn of illusions and excuses, they longed for a leader to guide them back to the right path. They were prepared to give up the weaknesses and follies of the past, to make a new beginning as a cohesive, caring people ready to work together again to achieve the vision that had created their homeland 25 years ago. All they needed was a leader who felt the same pain and yearned for the same goal.

Zulfiqar Ali Bhutto could have been that man. He could have rallied the people of Pakistan, led them to reverse the decline of the past many years and recreate the nation that had, against all odds, established Pakistan in 1947. But at this critical juncture, the man of the hour was found pitifully wanting. His lack of vision, meanness of spirit, and pettiness of mind, all led him to see this historic moment as just an opportunity to grab personal power. His limitations were exposed in the uses to which he put this power: the arrest and public humiliation of persons against whom he harboured personal grudges; the childish revelling in the trappings of office; the viciousness with which he treated his political opponents, and even some long-time supporters who did not show him due deference.

Sick at heart, almost as a diversion, I began to take an interest in Aleem's reports about his contacts with military officers who were talking about overthrowing the Bhutto regime. I suppose deep down there was a secret hope that perhaps I could help undo what I had done. It was quite clear to me that the young officers in the army and air force with whom Aleem was in touch could not, by themselves and with what they had available, achieve what they wanted to do. Such a goal would only be achievable if some senior officers in command of sizable bodies of troops were involved. Over these sterile months I spoke to a few friends, some serving, others retired. I also paid a few visits to Lahore and Rawalpindi, and met with some of the officers Aleem was talking to.

This was a time of emptiness, of great bleakness, in my life. I was lost, adrift, with no purpose. I could not see where I was headed, or what I could do that would be worthwhile. I went through the motions of living every day, but as an empty shell, a robot. Almost from the days when I first became self-aware, I had had some aim, some goal, in sight to which I could devote my life. Now I had none. Bereft, I tried, and mostly failed, to summon up some enthusiasm for the budding 'conspiracy' that was still mostly talk. For lack of anything better to do, I kept up my desultory efforts to find senior officers who might be willing to participate.

Appendix 1

B. Y. Ali

My father, Bakht Yawar Ali, or B.Y. Ali (as he was commonly known), was a remarkable person in many ways. I write about him here because much of what I am (and have been during the course of my life) is due to him, to what I inherited from him and the upbringing he gave me (and my brothers). I had, and have, a deep and abiding love for him.

He was born to a humble family in Patiala, a princely state in British India (his father was a small-time farmer). He was orphaned at a very young age and brought up by his two eldest brothers. They were quite poor; he once told me that, as a child, he used to walk barefoot to school as he had no shoes, which was especially hard in summer when the road was hot.

My father was very intelligent, and good in his studies. After graduating from the local college, through sheer merit he obtained a highly coveted admission to Government College, Lahore, the premier educational institution in the Punjab, to do his Masters. After completing this, he sat for the countrywide competitive examination held by the Government of India for the Indian Police. He was selected and, after training college, joined the police service in 1924. Most of his service was spent as the police chief in various districts of the Punjab, but also included a posting as head of the provincial railway police. In 1955 he retired as Director of Civil Defence for the province.

In addition to his intelligence, he had many other notable qualities:

Love and Gentleness: My father had a deep wellspring of love in his heart, which he lavished on his family, and especially on us, his children. He was not demonstrative at all but exuded this emotion as an aura around him that engulfed one in a warm glow. For others it was expressed as a pervasive gentleness in his dealings with them. I don't think I ever saw him express anger with anyone. If he was ever displeased by something we children did, the greatest reproof we got was a frown (of course, its impact on us was as strong as a loud upbraiding).

Simplicity: He was a simple man, with simple tastes and simple pleasures. He found joy in spending time at home with his family. He was fond of reading, and always had a book going. This simplicity governed his modest lifestyle, shorn of all extravagance.

Fortitude and Equanimity: Fate was not kind to my father: he lost a son in the prime of life. In his career, he was cheated of the success he was capable of and deserved; his nature was not suited to the conditions required for advancement in those times. During the days of British rule he did not emulate those ambitious local officials who aped the manners and modes of their foreign masters ("brown Englishmen", they were derisively called by other Indians). In Pakistan he would not stoop to the servility and sleazy tricks that advancement required. He had to witness persons junior to him (and less able) pass him by and move ahead. The hopes and ambitions with which he must have started his career never came to full fruition. Yet he never showed any signs of disappointment and despondency, or frustration and anger. He took it all in stride, and whatever burden he carried on this account he never displayed.

My father (1900–1960)

Kaiser (1930–1952)

The death of my brother, Kaiser, was a grievous blow, but he showed the deep grief he felt only for a day or two, and then went on with the same fortitude and equanimity that marked his whole life. But that grief, so rare and poignant, pierced my heart; it is a wound I bear to this day. As I recorded soon after: *That night, as we lay in our beds in the dark, he spoke of him. From the sound of his voice I could make out that he was also crying. It broke me completely; I could not bear the grief in his words and in his voice. From my earliest memories, he had been for me a model of courage and fortitude. Life's problems and worries had never ruffled his composure; in fact, the stream of life seemed to flow around his feet while he stood above it, calm, dignified, always in good humour, loving, lovable – a mighty tower of strength.*

Courage**:** He was a brave man. He had not only the moral courage with which he faced and stared down the vicissitudes of his life, but also the physical courage to confront extreme danger. He was awarded a gallantry medal for just such an act. He was the district head of police when he received a message one day that a policeman at the police barracks had shot and killed the commander of the barracks. When he arrived there the man was standing with his rifle near the body of his victim while all around people were hiding, afraid to confront him or even expose themselves. My father walked forward and ordered the policeman to lay down his weapon. The man was quiet for a few moments, and then said, "Come and take it". My father walked up to him and took the rifle from his hands. The sheer guts required to take those two dozen steps can be imagined.

He was indeed a remarkable man. His heart of oak carried him through life, serene and poised, absorbing the many blows that struck him, until it finally could go on no longer, and fluttered to a stop. He died in an instant, calm and peaceful as he had always been.

Appendix 2

Brief to the Hamood-ur-Rahman Commission

From: Brigadier F.B. Ali
Commander Artillery
6 Armoured Division

To: The Commission of Inquiry
c/o Supreme Court of Pakistan
Lahore

Sirs,

In response to your notification published in the press I have the honour to submit the following for your consideration.

I have 24 years of service in the Pakistan Army, and have received the best possible professional training and experience that the Army can provide, which I have supplemented by a deep and wide-ranging study of the subject of warfare. I can thus lay claim to being an expert in this field.

As Commander Artillery of 6 Armoured Division and of Army Reserve North during the war I was in a position to observe at close quarters much of the planning and conduct of the war in the sector of 1 Corps, which operated in the Sialkot–

Shakargarh–Narowal area. I am thus an eyewitness to many of the events which have a bearing on the cease-fire in West Pakistan.

I am in a position to, and am prepared to, appear before the Commission both as a witness to facts and as an expert witness relating to the matters being enquired into by the Commission.

I have the following points to submit:

a. The major reason for the debacle in both East and West Pakistan is that we did not have a war plan which was based on objective reality, and which would meet the requirements of the strategic situation. Specifically:

(1) It should have been clear to our political and military planners that, in case of full-scale war with India, our forces in East Pakistan could not hold out for very long against the superior might that India could mass against them. It should, therefore, have been obvious to them that the succour of these forces, in fact the very retention of East Pakistan, depended on our military response in West Pakistan.

(2) It should also have been apparent that the only military action in West Pakistan that could have affected the battle in East Pakistan was a massive offensive with all our available forces which, either by the seizure of vital Indian territory or by the destruction of substantial portions of his forces, would have persuaded him to accept an overall cease-fire. If he failed to agree to this, operations could have been continued to cause him such losses that his gains in East Pakistan would have been nullified.

(3) Even though we initiated operations in West Pakistan on 3 Dec 71 we did not launch anything resembling the big offensive that the realities of the military situation demanded.

(4) It is my considered opinion, and that of other experts, that we had the means and the opportunity of launching such a large-scale offensive on 3 or 4 Dec 71, and that the strategic and tactical environment offered good chances of achieving the desired results. Our failure to do this can fairly be regarded as a decisive blunder.

b. The war plan actually executed in West Pakistan was wrong and fallacious. This plan had been formulated quite some time ago on the basis that the Indian Army's main strategic theatre would be in the West, and, therefore, they would open the war with a major offensive in this theatre. Since about June 1971 it was clear that India's war aims, and hence strategic objectives, had changed. Their main theatre had become the East, and it was there that they would launch their main military effort. It was unlikely in the extreme that they would, at the same time, launch a major offensive in the West. On the other hand, it was logical to assume that in all probability they would remain on the defensive in the West until they had completed their conquest of East Pakistan. But we do not appear to have changed our war plan to fit the changed strategic situation. Instead, the Army was deployed in West Pakistan to receive and contain a major Indian offensive, which in fact never came. This deep, defensive deployment enabled the enemy to seize cheaply large tracts of land in our border areas, which he could not have done if a correct deployment had taken place. This easy gain of territory, and consequent improvement in his strategic posture, enabled him to commence fairly sizable offensive operations in 1 Corps area (from about 15 Dec onwards) and in the Rajastan sector. These losses of territory, and the commanders' lack of confidence in their ability to hold enemy attacks, were a major cause of our acceptance of the cease-fire in West Pakistan. But the basic cause remains the faulty war plan on which we tried to fight.

c. Another major reason for the debacle in West Pakistan was the general incompetence displayed in the conduct of the war by the Military Operations Directorate and other higher authorities at GHQ. From my viewpoint it appeared that they displayed a depressing lack of ability, foresight, imagination and even plain common sense, besides being unresponsive to warnings and suggestions from lower headquarters. To quote one instance, from about 12 Dec onwards HQ Army Reserve North kept continuously warning them of the deterioration of the situation in 8 Division sector, and urging that 6 Armoured Division be moved closer to this sector, and its artillery, which was lying idle, be allowed to support 8 Division. All these requests were refused, with the assurance that 48 hours' notice would be given before the Armoured Division would be required to move. The last such request refused and reassurance given was at 3 p.m. on 16 Dec 71. At 6:30 p.m. that same evening GHQ ordered the Armoured Division to move forthwith to 8 Division's area. Only a military man can savour the full flavour of folly and incompetence that this order contains.

d. Another major reason for the failure in West Pakistan was the utter incompetence displayed by Commander 1 Corps, his staff, and many of his subordinate commanders. His handling of his formation during the entire war is a sorry tale unrelieved by any bright spot. Some of the specific items, chosen at random:

(1) Under his orders, or those of his subordinates, large tracts of our territory were unnecessarily abandoned to the enemy, often without firing a shot. He did not make a single attempt to recover any of this territory, even though he had both the means and the opportunity.

(2) He failed to make any plans to use his Corps reserve, a whole armoured brigade, and in the event frittered it away in panic moves to block enemy advances. On 12 Dec 71, in my presence, the ex-Chief of Staff, Gen. A. Hamid Khan, gave Commander 1 Corps an order to use his reserve to make a flank attack on the enemy from north of Zafarwal. This was never carried out.

(3) He failed to make proper use of all the artillery available to him which, in my opinion, could have radically altered the situation in his sector. This in spite of repeated urgings from artillery advisors. On 12 Dec, another order given to him in my presence by Gen. A. Hamid Khan was to concentrate his Corps Artillery, and use it as such. This, too, was never carried out. The potential available is clearly spelt out in an official letter I wrote on 15 Dec 71 to the Commander, Army Reserve North, a copy of which I attach.

(4) 15 Division, a part of 1 Corps, took practically no part in the entire war, apart from abandoning the Phulkian salient, an area of considerable military significance, almost without firing a shot.

(5) At least two units of 1 Corps, 13 Lancers and 35 FF, were needlessly sacrificed, and suffered severe casualties. Even average military commanders would have avoided recourse to such desperate gambles.

(6) Commander 1 Corps displayed such concern and anxiety for his own personal safety as to seriously weaken the confidence of his subordinates in him.

e. Yet another main reason for the cease-fire in West Pakistan was the serious military situation in which our forces found themselves in the Rajastan sector. I can speak with some authority on this subject because from May 1970 to May 1971 I was the Colonel Staff of 18 Division, responsible for this sector, and am fully familiar with its operational area and plans. When I left this Division it had a sound operational plan which would have adequately coped with the enemy actions that occurred. It appears that the Division abandoned this plan and instead launched a hare-brained attack in impossible terrain. This met a severe reverse. Anyone familiar with the area and the limitations it imposed could easily have forecast this outcome. Though this Division commander has since been retired, the accountability of those who permitted him to launch this mad venture has not yet been brought up. It also appears that the plans for the Khokhropar sector were not properly carried out, since they would not have permitted the loss of the large territory which we, in fact, did lose.

f. A last cause for our acceptance of the cease-fire that I am aware of is the serious damage inflicted by the enemy Air Force on our strategic communications, especially railways. This danger could have been foreseen and, in fact, was by many officers who have been urging for long the need to improve our anti-aircraft defences. But no heed was paid to them.

As I submitted earlier, I would like to have an opportunity to appear before the Commission and testify to the above matters, and other related ones. If there is some way in which I can assist the Commission in arriving at the truth, I would gladly offer my services.

I am, Sirs,

Your obedient servant,

Sd / x x x

Brigadier
(F.B. Ali)

Attachment to Brief

IMMEDIATE
HQ Artillery
Army Reserve North
Operations Area
G-0113/10
15 Dec 71

To: Maj Gen Muhammad Bashir Khan
Commander Army Reserve North

Info: Maj Gen M I Karim, SQA, T Pk
GOC 6 Armoured Division

Maj Gen R D Shamim
GOC 17 Division

Subject: **Employment of Artillery**

1. I would like to submit for your attention the fact that the Artillery is NOT being properly employed at present. It is also possible that its full potential is not being given due weight in planning future operations.
2. As your Commander Artillery I have the means available to mass 12 to 14 regiments of artillery, ranging in calibre from field to heavy, on any sector you choose. The guns have been allotted, ammunition allocated, plans and preparations made,

and preliminary orders issued. Barring a few minor deficiencies to be made up in the command and control set-up, the only thing required to achieve this concentration of artillery is the decision as to where and when.

3. As your Artillery Advisor I would like to submit that, with this massive concentration of fire, great operational possibilities are open, provided there is enough infantry and armour to make full use of it according to a well-knit plan.
4. In case it is not intended to use this potential mass of artillery in support of Army Reserve operations in the next few days, I would strongly recommend that this mass be created and used by 1 Corps to recover some of the territory seized by the enemy. Needless to say, during this utilization, artillery liaison and advice will be available at practically all levels of the Army Reserve for planning any offensive operations you have in mind.
5. In conclusion, I submit that it would be most regrettable if the war were to end without us letting the enemy feel the full weight of this artillery potential.

Sd / x x x

Brigadier
Commander
(F. B. Ali)

PART 2

THE JOURNEY

6

The Journey Begins

I knew it was probably all up when I met Mirza Hamdani[5] on the evening of 22 March 73 in Rawalpindi; he told me he had met an old colleague of his who was in the Intelligence and he had told him they were aware of a plot brewing in the military against the government. There was a possibility the trail had not led to me, but it was a slim one. Any remaining hope vanished when I returned to Karachi and went to my office on the 26th; someone from Military Intelligence had been snooping around the building making enquiries about me. I knew then it was all up. I wrote at once (in veiled language) to Aleem Afridi telling him about this.

The next two days were difficult ones, filled with a horrible feeling of the net closing around me. I could see clearly what would happen. I also knew that I could, if I so chose, escape: I could go underground or I could leave the country. I discussed it with Jamila, but she was not much help; as she said later, she didn't really believe I could be arrested. I needed help, and there was no one I could go to, no one with whom I could even discuss it. I would have done so with Ijaz but he was away.

---

[5] Lt Col Muzaffar Hussain Hamdani, Artillery.

There were many problems with disappearing. One was to be absolutely sure they were going to arrest me, since there was the possibility they only suspected me; disappearing would be an admission of guilt. Then, of course, there were the mechanics of the thing. And, finally, there was the likelihood that if I disappeared they would go on harassing the family until I gave myself up. I agonized over these problems for two days, and then finally decided it wasn't worth it. If I was to be arrested, then let it be so. It was a great relief to end the indecision; I was at peace again.

On the evening of the 27th or next morning I learnt that Bhutto was to make a radio/TV speech on the 29th. It was likely we would be arrested the previous night. I had still not finally decided to close off all my options, and it occurred to me I could sleep somewhere else on the 28th just to see if they came for me. I considered some possibilities, and even went to see a friend, but (witnessing his reaction) ultimately gave up the idea. I tidied up my affairs, withdrew the little money I had in the bank, and discussed with Jamila what she should do after my arrest. That night I went to sleep expecting to be woken at about 2 (I don't know why, but I had the idea that was the time they always came for you). But no one came. I remember how pleasant it felt to wake up next morning and find myself safe and sound in my bed.

Since the 26th I had been coming home for lunch and not going back to the office. On the 29th I did the same and spent the afternoon at home. That evening I sat working on sketches for the architect of the house we had been planning to have built. At about 7 or 7.30 Bhutto came on the TV and we sat and listened to his speech. The usual rambling affair, it went on for almost 2 hours. The only pertinent reference seemed to be where he said he would smash any conspiracies against him. It was fascinating sitting there watching him, as if looking close up at a deadly cobra, its hood spread in evil menace. (And all the time with the bitter taste of defeat in my mouth).

Soon after 11 p.m. the bell rang. It was Kafil at the gate (he was a PMA course-mate of mine). He called me outside and, walking up and down the lane, told me that someone in the police's Special Branch had rung him up wanting to know my address. He had told them he didn't know. He asked what it was all about; I told him I had no idea. He suggested it might be a good thing if I spent the night somewhere else. I replied that it didn't really matter.

I had hardly finished telling Jamila about Kafil when the bell rang again. This time it was Anwar and Kaukab (Jamila's sister and brother-in-law). They had received a call from Iftikhar, her brother in Rawalpindi, who told them he'd been rung up by a Major Bajwa of the Coast Guards from Karachi who wanted to know my address.

The reason the Intelligence was frantically trying to find out where I lived was because I had given them the wrong address! A couple of days earlier Jamila had asked me to complain to the dairy farm manager about some milk delivery problem; when I spoke to him on the phone he asked me for my address; knowing that my line was being tapped I gave him a wrong one. Anyway, they had the address now, and tonight was the night. While we stood talking in the drive (the Makhdums did not come inside so as not to alarm Jamila's mother who was staying with us) the ISI staked out the house. Two of them started strolling up and down in the lane outside, and later I saw others at several crossings around the house.

Jamila wanted us to go and see Manzur Qadir (probably Pakistan's premier lawyer, whom we had met through M. Anwar in connection with our planned court case). We had read in the papers that he was visiting Karachi, and I knew he always stayed at the Sind Club. It was about midnight now, and I didn't want to go as I thought he would be asleep. But Jamila insisted, and so we went. No one followed us, even though they had a motorcycle hidden in a side-lane.

There was a light in Manzur Qadir's room and, as we knocked, Tajammul Hussain came out to leave. In a few more minutes MQ, who received us in his night clothes, would have gone to bed, and then I would certainly not have roused him. We told him our problem. First he said that it may be just a scare, the Government sometimes did that; I told him this was not one of those cases. He then said I could exercise a choice of where I wanted to be arrested; I replied Karachi was as good a place as any other. He suggested we ensure that someone followed me when they took me away so it was known where I had been taken. Then Jamila should come and see him in the morning. He was very kind and helpful. He joked that it seemed to be a great day for arrests, as we were the third party who had approached him with the same problem that day. He tried to console us by saying that in Pakistan it was no longer a shame to be arrested, as the best people seemed to be meeting this fate. When we left, his son, Bisharat, saw us to the car; he was also very sympathetic. It was a great load off my mind because now Jamila knew exactly what to do, and would be in the best of hands. She was also feeling much more reassured.

When we returned to the house there were even more people around, and also a tell-tale Volkswagen in the side-lane. We decided that Anwar should spend the night at our place so he could follow when they took me away. When he went to drop Kaukab home I went with them; again no one followed us. It was past 1 when we went to bed, and I lay awake waiting for the expected 2 o'clock ring. After a while I dropped off to sleep.

I awoke with a curious sound of rustling in my ears. I hardly had time to realize this was the sound of many feet on the dry leaves in our compound, when a long ring on the buzzer shattered the peace of the night. Actually, it was 5 a.m. though still dark, and a large force of police had surrounded the house.

We were all up at once, and I told Yawar to go to the front door while I splashed some water on my face. Just then they started banging on the door, and shouting "Kholo! Kholo!" I quickly dried my face, and walked down the passage towards the front door; Jamila had gone a little ahead of me. Just then Afroze opened her door, which overlooked the lobby.

When I got to the lobby I found the door half open and quite a collection of people outside, mostly police officers though the person holding the door open was in civilian clothes. It seems that when Yawar had opened the door they had barged in but Jamila had ordered them out again. As I came into view, the civilian asked me "Are you Brig FB Ali?" When I said I was, he said, "Sir, I am Mueen-ud-din, SSP Karachi, and I have a warrant for your arrest and for the search of your house". At this Afroze let out a little scream of terror, "Mommy!" We had not told the children anything.

I asked them to come in, and took them into the drawing room. When they were settling down, Lt Col Faghfoor (local head of the ISI) came in from behind; he looked very sheepish and started apologizing. I asked him what was the need for all this drama; he could have asked me to come to his office and arrested me there. Shortly after, he went out and stayed outside throughout. There was another person in civilian clothes who was introduced as Major Bajwa of the ISI.

First of all I read the warrant. It was a handwritten note of the District Magistrate, Karachi, on a handwritten letter of the SSP (Senior Superintendent of Police) to him saying he had received a telephonic request from the SSP Rawalpindi for my arrest and house search as a case had been registered against me under various sections of the penal code. I copied out these sections for Jamila to show Manzur Qadir. The SSP suggested that while I packed, his officers would search the house. He also advised my wife to put in some jerseys as it would be chilly where I was going

I shaved and dressed while Jamila packed, and the police officers searched the house. Afroze had recovered from her initial shock and went around making flippant and sarcastic comments to them; Yawar was his usual quiet self. Anwar appeared completely stunned and just sat on his bed. Jamila's mother (who was partially paralysed after a stroke) was in one of the rooms and we wanted to avoid alarming her. Jamila told her the police had come to carry out a routine search for arms; I don't know how much she believed, but she just lay quietly in her room. I think her attendant slept right through the whole thing!

By the time I had got ready it was past 6, and the SSP was becoming nervous as it was getting light outside. He kept reassuring my wife I would be looked after. Finally, I said I was ready to go, and then there was some delay as their vehicles had to be brought from wherever they had hidden them. So, for a while we stood together on the porch – Jamila, Yawar and I (Afroze, I think, was still inside where the search was continuing). It was cool and pleasant and quiet; the world had not quite got up yet. We did not say much to each other; there was very little to say. Now that it had happened, I was keen to be off to whatever lay in store for me.

Faghfoor's white Mercedes drove up. I said goodbye to my family and got in behind with the SSP while Faghfoor sat up front with the driver. We drove off at once, going fast. Jamila was supposed to follow with Anwar but I could see that they were going to be delayed.

There wasn't too much traffic on Drigh Road. Looking around it appeared strange to see the world going about its daily business. The SSP remarked that it seemed as if I had been expecting them. I told him I had been expecting this ever since my retirement. He asked me some questions about the events of those December days when Yahya fell, and which had led to my being retired from the army.

They took me to the Clifton Police Station. From there the SSP rang someone to tell him that he had brought me there. He spoke to Faghfoor about getting me breakfast from the Metropole hotel! I was put in a room with a guard outside. They brought me some tea and a newspaper. I listened for any indication that Jamila had come. Once I thought I heard something, but there was nothing further. Knowing she had got left behind I had asked Bajwa to see that she was told where I was; he promised to send a car.

I must have been there for about 45 minutes when Bajwa arrived and asked me to come along. He seemed in a hurry. We got into his station wagon with a police official up front and drove to the PAF Drigh Road airfield. I kept a lookout for Jamila but drew a blank. I also kept looking around at this familiar route, which it seemed I was traversing for the last time. It was still early, still not too noisy; a quiet, cool morning, strangely peaceful.

Bajwa seemed very upset. He saw that I was quite calm, and thought perhaps I did not realize how serious my situation was. He kept repeating this to me. He also kept saying how sorry he was, and expressing bewilderment at what had happened. He had seen me when I was the Colonel Staff in the division HQ there, and I suppose the contrast shook him. He told me how everyone used to say I would certainly become a general one day.

We stopped at the level crossing just before the airfield gate. Here he told me that when we arrived the police chap would handcuff me; those were the orders. (Here it starts, I said to myself). We arrived at the air terminal building where there was a PAF Group Captain (I learnt later he was their Provost Marshal). I don't think he quite realized who I was, because after shaking hands he rather hurriedly disappeared inside with Bajwa. The SI apologetically handcuffed me, putting both cuffs on my left wrist. We then sat in the verandah. A C-130 plane stood nearby.

Soon a van arrived and drove up next to the plane; I saw Hashmi[6] taken out and led on to the plane. He was in his nightclothes, handcuffed and blindfolded. Seeing that the trail had led to Hashmi I again considered who had put the finger on us. At that moment Ghaus seemed a good bet.

Bajwa came out saying he had just been ordered (on the phone) to accompany me; he also said he had been ordered to blindfold me. He got a cloth from somewhere and did this. I was then led onto the plane, and it took off. I asked Bajwa to remove the blindfold but, after checking, he told me the Air Force officers did not agree. After a while he brought a breakfast pack and helped me to eat it.

It was a lonely journey, sitting there handcuffed and blindfolded. Leaving behind those I loved to an uncertain fate, and going into an unknown future that promised to be anything but pleasant. Manzur Qadir had indicated that moving a person after arrest to another province required a certain procedure, took time, and could be fought in the courts. But here I was, within an hour of my arrest handed over to the Army and flown out to an unknown destination. Throughout the flight I kept steeling myself to undergo

---

[6] Wing Commander Hashmi, PAF. Sq Ldr Ghaus was the air force officer who had initially contacted Aleem Afridi.

what I knew was to come; the attitude displayed so far promised that things were going to be rough. Underneath it all, a feeling of utter desolation. Everything was finished.

We must have flown for 3 to 4 hours. Once I told Bajwa, who had been sitting next to me, that I wanted to go to the toilet, but apparently someone else had taken Bajwa's place, and he replied with a giggle that there wasn't any on the plane.

Finally the plane landed. The door opened and a gush of hot air hit me. I heard several vehicles drive up to the plane. I was led out to what appeared to be a van. I think Bajwa asked whoever was in charge about allowing me to urinate, but he refused. Some other people got in and we drove off. Soon we were driving through crowded roads, with many tongas (still blindfolded, I had to depend on what I heard). I thought first it was Multan, and I was being taken to the fort there, but we kept driving on and I ruled that out. Then I guessed that it must be Rawalpindi (I discovered later it was actually Peshawar, presumably because that was where the air force wanted to take Wing Comd Hashmi, my fellow prisoner on the flight).

We were out of the town and the convoy speeded up. Our van's front wheels started to wobble at that speed. After a while we stopped; I was taken out and transferred to a jeep. While I was being moved, blindfolded, one or two buses passed – they must have wondered what was going on. The jeep moved off. It was a strange feeling hearing the two (or was it three?) officers chatting away as people do who have met after some time, occasionally laughing, apparently totally unaware of me sitting in the back of the jeep, blind and handcuffed, amidst the ruins of my world that had collapsed upon me that morning.

7

Attock Fort

(March 1973 to March 1974)

After about an hour the road started to twist and turn and I guessed we had come to Attock Fort. We drove into the Fort, stopped, and I was led out. Someone asked me if I wanted to go to the toilet, and I was taken to one. Once inside, my blindfold was removed and I saw that holding the chain on my handcuffs was a plump young Major looking rather embarrassed. Afterwards, the blindfold was again put on and I was led to a room, where it was taken off. It appeared to be an office, and my escort and I sat down on a couple of adjoining chairs. A hefty Air Force MP came and stood at the door. My suitcase was also brought in and I took out a cigarette and smoked it.

After a while a very harassed Captain came in to record my number, rank and name (as usual in the army, he was making a list!). Then Bajwa turned up. He was still overwhelmed by the enormity of what had happened, and puzzled at my apparent calm. He said, "*Sir, it's something very serious. You don't realize how serious it is .... I am terribly sorry this has happened. I keep asking myself why God should have done this to you. The only thing I can think of is that when I met you as Colonel Staff, I thought you were quite proud. I suppose people who are self-respecting are also proud...... Anyway, I admire the way in which you and your family have taken this*". He said he would shortly be going back. I asked him not to forget to take his handcuffs with him.

After some time I was again blindfolded, and Bajwa took me to a jeep. We drove a very short distance, then Bajwa helped me up a fairly long flight of stairs into a room where I was made to sit on a charpoy. Someone said my room was being got ready and I would soon be taken there. Bajwa took off his handcuffs and someone put another pair on. Then I heard Bajwa say goodbye (still blindfolded, I could not see him). Turning in his direction I thanked him for the kindness he had shown; I also said to him, "*Bajwa, it was never pride. Reserve, yes, but pride, never.*" Bajwa's departure left me feeling all the more alone. He was a link to Karachi; he was also sympathetic and considerate. I did not know in whose hands he was leaving me.

Shortly, I was taken to another room down the verandah. The blindfold and handcuffs were removed. The iron-barred door clanged shut, the key was turned in the lock, and a sentry with his rifle stood outside the door, looking in. I found myself in a large high-ceilinged room, with a floor of large slate pieces, and dirty, patched walls. In the centre was a crooked 'baan' charpoy with a filthy mattress and pillow and a couple of typical army-issue blankets, stiff with dirt. The time was probably about 2 p.m.

From the door of my cell I could see part of the inside of the Fort. There was great activity, with trucks coming and going, unloading cement, sand, bricks etc, and what appeared to be an engineer unit at work. Directly in front of me, but lower down, a few officers sat under a tree, reading newspapers and drinking tea. They seemed to me beings of a different species.

I wondered if there was anything in the newspaper about the arrests. And how the news had got to my mother and my brother, Dani; and how they had taken it. And the consternation it would cause to all who knew me and cared. And who else had been arrested with me. After a while I asked a young officer rushing about in the verandah whether they provided any lunch there. He said it was being arranged. After some time lunch arrived: some 'dal' in a tin plate and a couple of 'rotis', all dumped in my lap. I ate a few mouthfuls.

Later two Lt Cols in uniform came into my room. One was a big, fat, tall chap with white hair, who I later on came to know as Aziz, the Fort commandant. I had seen them both walking up and down outside, and it seemed Aziz was the senior chap in the place. The moment he came in I confronted him: *Why have I been brought here? I am a civilian and cannot be kept in military custody. I want to see my lawyer.* He was rather apologetic, saying he knew nothing about anything, and would pass my message on to those responsible. He then carried out a perfunctory search, spending a while examining my shoes.

I spent the afternoon walking about my cell, looking out the door, trying to get my bearings and learn as much as I could about the set up. Late in the afternoon I saw another prisoner brought in, the jeep stopping right in front of my barrack. A tall, well-built man, in 'shalwar kameez', blindfolded and handcuffed, was led up the many steps, and then past my door into the cell next to mine. I didn't know him, but later learnt he was Major Nasrullah Khan. He sighed deeply as he went past my door.

In the evening it drizzled for a while and became rather chilly. The bulb in my room wasn't very bright. My suitcase had not been given to me, and later I asked for it to get my night clothes. It was brought outside the door, and putting my hands through the bars I took out some clothes. I have forgotten what dinner was like; probably the same as lunch. After eating a little I changed under the eyes of the guard outside the door and went to bed. The light burned overhead all night, and I had to cover my eyes with the sheet to sleep. The first day was done.

Next morning there was the problem of getting ready. I asked for my toilet things from my suitcase. There was some delay as I think they called Lt Col Munir before giving them to me. When I asked for a cigarette it was refused. I had to ask Munir, who told them to give me one. A commode was brought in and placed in a corner out of sight of the door; it wasn't too clean. This was the only one available for a couple of barracks, and it was apparently being moved from room to room!

After I got ready I spent most of the time walking about my cell and looking out. The building activity went on at full swing; they appeared to be converting barracks into cells. Sometime in the morning a funny looking Brigadier came to my cell-door. Later I came to know him as Ghouse, commandant of the nearby AK Regimental Centre and now also overseeing the Fort prison. I protested to him about my detention, but he too expressed ignorance of the affair, and said the people concerned would be arriving soon.

A little later I saw some officers in civilian clothes knocking around, and guessed that they must be the interrogators. Later that day a short pudgy officer, also in civilian clothes, with a cigar in his mouth, came around inspecting the inmates. He stopped outside my cell-door, and stood there for a while, cigar in mouth, smiling. I looked at him once and then ignored him. His face seemed familiar, and later I recollected that he was Anjum, whom I had once met while visiting GHQ as the Colonel Staff of 18 Division; he was then a Lt Col in MI Directorate.

That time he had been running around at my beck and call, now he was a Brigadier and in-charge of the interrogation.

****

(*Recounting this gloomy tale so many years later, it seems appropriate to lighten the narrative somewhat by adding the tale of another arrival at Attock Fort*).

Major Sajjad Akbar (Catlow, as he was known to his friends) was working as a staff officer at GHQ. A friend of Nadir Parvez, he must have been in some gatherings where these young officers let off steam. A day or so after I was arrested Sajjad was called in by his Director, who said, *I'm sorry, but I've been asked to have you sent to Attock Fort as it seems you were involved in a coup attempt.* Sajjad digested this for a few moments, then, genuinely puzzled, asked, *Sir, what is a coup?* The Brigadier explained what the word meant, then called in a couple of officers waiting outside and handed Sajjad over to them.

By the time they arrived at Attock it was past noon so they decided, before going into the Fort, to first have lunch in the newly set up mess for the officers of the guard contingent. While they were having lunch another group of officers arrived and sat down at the other end of the table. They appeared to be new arrivals who obviously hadn't yet been told what was going on and who were these prisoners being brought there whom they were to guard. They began an excited discussion about this, with various theories being advanced.

By then, Sajjad and his escort had finished lunch, and as they were walking out Sajjad stopped near this group, still in animated discussion, and said, *I know.* There was a sudden hush and everyone turned to look expectantly at him. Sajjad paused for a moment, then gravely said, *Coup*! and walked out, leaving several bewildered faces peering at his back.

As they entered the Fort gate they met a group of NCOs belonging to the 'rear party' of the SSG (commando) battalion that had been based in the Fort but was now deployed elsewhere. Sajjad was a former SSG officer and these men greeted him effusively. One of them asked him, *Sajjad sahib, have you also brought some murghas*? (using the Punjabi term for a trussed-up chicken, and referring to the cuffed and blindfolded prisoners being brought to the Fort). Sajjad just smiled and walked on.

The party walked to the Fort office which Sajjad and the escort officers entered. The SSG men waiting outside were stunned when, a short while later, Sajjad appeared, handcuffed and blindfolded (a right proper *murgha*), being led to his cell under guard.

****

On the third day they took away the broken-down bed and rotten bedding, and gave me a hospital bed with a thick foam mattress, clean sheets, new blankets and a good pillow. They also put a commode in a corner of the room. This felt like luxury, especially the bed. Soon after, they provided a straight backed chair and small bedside table with a drawer. However, my suitcase and personal effects were not given to me. For a few days I had my shaving kit but then they took this away also and I had to ask for it every morning. There was nothing to read, nothing to write with. I tried to get them to give me a book I had brought in my suitcase, but they refused.

The broader implications of my predicament soon hit me. The strongest initial emotion was anger at myself for getting mixed up in what was obviously a shambles. For not having exercised caution commensurate with what I could lose. It wasn't that I had been unaware of the risks involved; I consciously accepted them on the basis that someone had to stick his neck out. What I hadn't properly considered was the havoc it could wreak upon my family, and this now deeply oppressed me. We had just settled down in Karachi and started a new life; it was now shattered. Jamila would have to give up the house and leave Karachi. Yawar was going to school there; what would happen to his schooling? And, what of Afroze, my mother and Dani, my younger brother?

There was also bitter regret. Regret at all that was lost with nothing to show for it. My life felt finished; there was nothing to look forward to. The one highlight – December 1971 – had also turned sour: it, too, seemed one big fiasco. I felt a deep anger at those who had caused this shambles by their crass stupidity, the casual manner in which they had gone around talking to people about their intentions, their insane confidence that they would not be exposed, the total lack of any precautions.

Prey to all these feelings, I was afflicted with a sharp restlessness. I would spend hours walking about my cell. At times a wayward train of thought would take me, unsuspecting, around a corner and leave me face to face with a memory or an image, and I would find myself gasping, wracked by an anguish that sometimes made me cry out. What made the emotional turmoil worse was that I had no way of expressing it; there was no one to talk to. Later in April they allowed us to us buy a writing pad and pen (to write a letter home). I felt an urgent need to use this writing material to express what I was feeling; the only appropriate medium seemed to be verse, and over the next few weeks I composed some poems. I was myself somewhat surprised at the depth of hurt they revealed. They are reproduced in Appendix 3 at page 205.

The interrogations began the first day. If I remember aright, Nasrullah was taken out from his cell next door, blindfolded and handcuffed, at about dinner time and had not returned by the time I went to sleep. Several times during the succeeding days and nights I heard the clinking of handcuffs which meant the escort had arrived to take someone away for interrogation. The steps would come up to my door, I would tense and say, *Here it comes*, and then they would go past. I was not taken for several days.

I had been tensed up for my first interrogation ever since arriving at the Fort and the suspense wore me down. I expected it to be a no holds barred affair, quite possibly including the third-degree. It was impossible not to wonder how much I could stand. I looked very carefully at the prisoners returning from interrogation but they appeared to be all right. As the days passed, little by little I found I was better prepared to face the interrogators; it was on the first day that I had been most vulnerable. That day I knew very little of what had happened, but gradually I acquired bits and pieces of information.

Early on I learnt from the young lieutenant of the guard (Chughtai) that Squadron Leader Ghaus was also in the Fort. I had suspected him of being the informer but this cleared him, though I kept open the possibility this might also be a trick. The second evening I saw Farooq Adam and Asif Shafi being brought, blindfolded and handcuffed, into a barrack near mine. Then one or two others whom I did not recognize. On the third or fourth day they started to close the wooden doors of the cells before taking anyone out. Even then there was a peephole in my door through which I could see a little; this they didn't close till almost a month later.

My first interrogation was on the fourth or fifth day. I was taken to an office and after a short while Brig Anjum came in and shook hands. He said he was no interrogator; all he wanted was for me to tell him all I knew. He was aware that I was not too deeply involved, that perhaps there was no case against me (a sudden lift of the heart here), that I should help them to root out this cancer from the Army, etc. I told him I didn't know what he was talking about. He smiled and said they knew everything; Wajid, for example, had told them a lot, including about my visit to him in February. This shook me but I still said I knew nothing. He smiled again and said: *Think it over, we'll talk again.*

I was called again after a couple of days, perhaps three. In between I'd learnt who had betrayed us. I overheard Nasrullah telling Munir Rafiq (who was in the next adjoining cell) that the interrogators seemed to have a complete transcript of one of their conversations and he suspected the "Signals Lt Col" of having taped it. So it was Tariq Rafi! I realised it was no use telling them again what I had told Anjum.

This time I was taken to another office where there was a Lt Col who introduced himself as Habib. He was very nice, offered me tea and a cigarette, and chatted for a while. I gave him an outline account of what I 'knew', pitching Tariq Rafi as the principal instigator. While I was talking he suggested that as I wrote well, why didn't I write up my account? I told him I wouldn't write anything or sign anything, and I would give no names. He tried to persuade me but soon gave up.

A few days later I was again taken to Habib; I think this was on 11 April. He read out what he had written based on what I had told him last time. It was very badly drafted, and he again asked me to draft it myself. I told him I wouldn't do that but could dictate the statement to him. He agreed, but before that asked me a few questions. The first one shook me pretty badly; it was completely unexpected. It meant either Aleem or Sarwar had talked, probably the latter. There was absolutely no need for either of them to have done so since the interrogators would not otherwise have known of this incident. If they had been told this, then there probably wasn't much they didn't know. It was very upsetting.

Even though Habib was always very nice, the strain during these sessions was considerable. I was always very keyed-up, watching every word I uttered, trying to out-think the interrogator, to guess how much he knew already and what he was after. Even though it wasn't very warm in those days, the tension made me perspire in these sessions. Still, I did manage to get a lot of information out of the interrogators, mainly from what they asked me.

In going over our predicament I became more and more convinced that some means must be found of conveying to Bhutto the truth of the matter, namely, that the real motivation of these young officers was to rid the army of the corrupt and incompetent generals who had caused our defeat in the 1971 war. I felt that if this was done perhaps he would see what a great tragedy it would be to punish them, since it was their love of country and army that had led to their being imprisoned here. I was sure the Army would convey a completely false picture to him. At first I thought of writing a letter to Bhutto, but it seemed very doubtful it would get to him. Then it struck me that if there was one thing he might well read it was my interrogation statement; I could put into it what I wanted to convey.

Next time I saw Habib I told him I had changed my mind and would write out my statement; he seemed quite pleased. I did this, and in it I included what I intended to convey to Bhutto. It was mainly an effort to persuade him to let these young officers go. My statement was typed out and I signed it.

A strange experience befell me in late April. During an interrogation Habib asked me some questions about my Kharian visit. This shook me since I hadn't for a moment thought they would know about it. Then, as I was walking back to my cell I saw, walking ahead of me, a person I recognized as Azam; Aziz was escorting him to a cell. I was totally unprepared to find Azam here; I could not imagine how they could have got on to him (it also meant that they had got to Iftikhar, Jamila's brother, a Lt Col serving in GHQ). It seemed they really did know every single thing and had got everyone even remotely connected to the main group. It depressed me terribly and I remained so for the rest of the day. In the evening, as was my habit, I pulled up a chair near my cell door and sat there watching through the bars as the evening turned into night.

I was in an agony of doubt, sorrow and helplessness. Everything seemed hopeless. It felt as if God had turned his face away from me, from us. Sitting there, watching the light fade away, I kept asking God why he had abandoned me; what had I done to be thus cast away. Why...why...why? What had we all done to deserve this? The wreckage of so many lives, many of them so young, crushed my heart. In this agony I asked God for a sign that he had not abandoned me: to show me a light between the two bars of the cell door which were directly in front of me.

I sat for a long time looking out through the bars at the sky, and at the back of my mind was the idea that the light I wanted could perhaps be a flash of lightning or an aeroplane's lights. But it was a cry of desperation; deep down I knew such things did not happen; God did not give signs like this.

The evening light slowly faded, it became dark. I must have been sitting for an hour or more when it suddenly happened. A bright light flared up in front of me, exactly where I was looking, between the two bars I had picked. An electric shock went through me; I half arose from the chair. For a few moments I could not understand what had happened, what this light was that had suddenly appeared in the sky. It went out, and then I realized what it was. I was facing the far wall of the Fort; someone had climbed to the top of the rampart and standing there had lit a cigarette. From where I sat the light of his match (perhaps two matches, for it was very bright) had appeared as if in the sky. But it was the sign I had been asking for – a light, and exactly where I had asked for it.

The first shock gave way to a feeling of exaltation: God had given me a sign. A sign that he had not abandoned me. Dimly I made out the white shape of the person who had lit the cigarette. For me at that moment he was not a man but a veritable angel of God. A feeling of great peace spread through me. Along with gratitude. The turmoil and tension fell away.

Afterwards I did ask myself if it really was a sign or just a coincidence. It was true that I had sat in the same spot every evening and not once seen anyone on the Fort wall opposite. And the light had appeared at the exact place I had picked. To put the matter to the test, for the next 15 evenings I sat in my doorway and watched the wall but what had happened that night was never repeated. Despite this questioning, in my heart I never really doubted, and from this day onwards I was at peace. I had no doubt everything would turn out well. This incident gave me great strength and fortitude all through these many difficult months.

Life in Attock slowly settled into a routine. I would wake at about 6.30 with the arrival of breakfast – an enamel mug of tea, a couple of fried eggs and two pieces of toast on an enamel plate with some butter and jam, a spoon, no knife or fork. After breakfast I would ask for a bucket of water. Because there was only one bucket in our block, sometimes I had to wait till it was available. When the water came I would wash and shave, using a corner of my cell as my bathroom. After I dressed the sweeper would come and sweep out the room, including the water that had collected in the corner, and take out the commode to clean. Sometimes he was busy elsewhere and one had to wait till he was free.

Then I would sit at my table and write a letter, or work at one of the poems, or just doodle. About 10.30 I got a mug of tea. This was one of the high points of the day. After the tea I asked for one of my cigarettes. They had allowed cigarettes after the first week, but the packet and matches were kept outside and each time you wanted one you had to ask the guard commander. Then I would either do some more writing, or walk about, or lie down, and sometimes even nap.

Lunch came at about 2 or 2.30. After lunch I had another cigarette, then would lie down and sometimes sleep. Tea arrived at about 4.30. After that I did some exercise, which consisted of walking round and round my room about 100 times (this equalled about a mile and a half) followed by some other exercises.

I would then wash, bring up a chair near the door, and sit there looking at the world outside. In front of me was the canteen where there was always quite a bit of activity with soldiers from the Fort garrison coming and going, or standing outside and chatting. I could also see some hills beyond the Fort walls, and every evening I watched the sun go down behind them. Dinner was brought at about 8. By about 9 I went to bed, wrapping the bed sheet around my head to dull the light that burned overhead all night. Usually I went off to sleep soon enough, though sometimes the music booming out of the canteen loudspeaker was a nuisance.

Thus passed day after day. The solitariness was almost complete. Apart from a word or two with the guards or an exchange of greetings with Chughtai, the young guard lieutenant, there was no conversation. I used to look forward to the daily visits of Boota, the sweeper, because with him I would exchange a few words. I found I had to talk to myself for company. Others used to sing or whistle, but I tried this only a few times. Then, I would spend long hours lying on the bed and trying to make out various shapes and forms on the walls. Some of these became old friends.

****

One day in the last week of April Chughtai whispered to me that a writ petition had been filed on my behalf in the Lahore High Court. A couple of days later he told me that I would soon be taken to Lahore to appear before the court. Early morning on the 29th I was taken to the Fort office and told I was to go to Lahore; I was handcuffed and made to sit in the back of a jeep with a couple of armed escorts while the young captain in-charge sat in front with the driver. However, as soon as we exited the Fort he stopped the jeep, had my handcuffs removed and asked me to sit up front with him while he drove the jeep. Throughout this trip of several days he was very respectful and helpful; I was touched by his consideration.

A feeling of exhilaration surged through me as we drove out of the Fort gates. In the cool of the early morning it was wonderful to be driving along thus on the highway. I looked with wonder at the passing cars and buses, at people going about their daily business, as they pleased. These everyday sights seemed strange and wondrous to me now, even though I had been imprisoned for just a month. I drank in the sights and sounds of this journey along a route I had traversed so many times before. When the sun got higher it became quite hot, and it was nice to stop in the shade to eat our packed lunches.

It was late afternoon when we reached the outskirts of Lahore. As we drove through these familiar places of my home town, excitement mounted within me and I eagerly looked around in the hope that I may see some familiar face among the passing scene. We drove through the city into the cantonment and stopped at an officers' mess. I was put up in the guest room and an armed guard was stationed outside. It seemed a great luxury to be in a properly furnished room with an attached bathroom. My contentment increased when they brought in my suitcase (I had not seen it since arriving at the Fort), and I eagerly searched for the book I had put in it in Karachi, only to feel a sharp pang of disappointment not to find it there – they had obviously removed it before putting the suitcase in the jeep.

Next morning, the 30th, as they drove me to the High Court, a feeling of panic arose within me at the thought of seeing the family again, so much so that for a few moments I even wished I was back in Attock! At the courts I was taken to one of the offices of the Advocate General to wait until I went to the courtroom. Suddenly, through the window, I saw them – Jamila, Afroze, my mother and Dani – in the small crowd that suddenly appeared outside (the court was apparently taking a break). My heart was pierced, tears welled up in my eyes, and I had to walk about in the room to conceal them from the escort officer. I feared I would not be able to control myself, that I would break down when we later met. They soon disappeared; the court must have resumed its session. After a while I was escorted to the courtroom.

They were all there, and many other relations besides. The room was full; the wives of some of the Attock officers were there, too, including Shamim Afridi (her husband, Aleem, had testified in the court before me). It was bewildering; looking at all of them, trying to get my fill of them before I was whisked away again, trying to grasp what was going on, all the while listening to the whispered questions and instructions. Then the judges (Shamim Hussain Kadri and Ch M Siddiq) came in, and I was called to the witness stand. M Anwar was our lawyer, and he had me describe the conditions under which I was being kept at Attock Fort and the treatment I was receiving. The Advocate General declined to cross-examine (even though the army's Judge Advocate General and some other officers were there to assist him) probably because my statement had been strictly factual. Before I was led away, the judges granted M Anwar's request that the family and he be allowed to meet me the next day.

That night, in my mess room, I wrote a letter laying out what I had gathered was the army's case against us. Next morning M Anwar arrived with Jamila, Afroze, my mother and brother, and we met in a mess room with an officer sitting in on the meeting. His presence naturally restricted the conversation we could have, besides causing me quite a bit of tension at the prospect of secretly passing on the letter I had written. Anwar cooked up some excuse to have the officer go out on some errand, and in a rather panicky way I slipped the letter to them (obviously even that one month in prison had taken its toll). They brought some books and a small battery radio which the court had permitted me to have. The stilted meeting ended all too soon, and they left saying that they would arrange another one for the next day (the officer assured them I would still be here).

As soon as I had finished lunch, an officer came and told me that I was being moved to another mess ("a better one") and I should quickly pack my things. In the jeep, with my bag and the escort, I soon realised that we were not heading to another mess but out towards the highway. The escort captain confirmed that his orders were to take me back to Attock. It was a long, sad, hot and forlorn journey back along the same road we had travelled only a couple of days ago. We arrived at the Fort in the late evening to find the commandant, Lt Col Aziz, waiting for us in a foul mood (apparently the newspapers had covered the court proceedings, including my statement and Aleem's, in some detail). He first chewed out the escort captain for having me sit in the front seat without any handcuffs, and said he would report him to his superiors. He then had some guards handcuff and blindfold me, and I was led in this state to my cell.

The writ petition had been about the conditions of our detention not being in accordance with military rules and regulations; the government did not challenge this and the court's comments had clearly shown they accepted this argument. So the army began to backtrack. The next day the books and radio I had brought from Lahore were given to me. It was sheer luxury being able to read after a whole month. As well, the joy of listening to the radio, of being able to hear the news and have some contact with the outside would. That day (May 2) they put ceiling fans in the rooms. After a few days we began to be taken out singly for walks inside the Fort. Being locked up, and cut off from the outside world by the Fort's walls, it was a great thing to get glimpses of the outside.

I remember when coming back from the interrogations I used to stop in the office verandah and look out at the magnificent view from there: the river flowing down below along the Fort wall, the rising ground on the other side that suddenly erupted into hills, the road and the railway; the houses, people, cars and buses, a world going about its normal humdrum everyday life. Later on, one evening I was walking up and down in front of the offices when I saw Saeed Akhtar Malik standing at the corner of this terrace looking out over this view. He stood there for a long time, and something in his manner made me think of a little boy standing in front of a shop window filled with the most gorgeous toys, looking at them longingly and wistfully, knowing that they could never be his; just standing there and looking and looking.

They started recording the Summary of Evidence (a necessary preliminary to the trial) sometime in May. This was the first time we were all brought together since our arrests. What a great feeling it was to be in human company again! I was meeting many of the other accused for the first time. After the weeks of silence there was a sudden outpouring of talk, which they vainly tried to stop. There were also the whispered consultations on the case, even though our search for hidden microphones had not turned up any. The principal witnesses against us came and had their statements recorded in our presence. The first was the one who had betrayed us, Lt Col Tariq Rafi.

As the evidence unfolded it soon became clear how we had been exposed. Tariq Rafi held a key position, commanding the GHQ Signals Battalion; he was also a friend of Asif Shafi's. Asif went to see him one evening and sounded him out; finding him receptive he told him about their thinking, and Rafi agreed to participate. When he was leaving, Rafi offered to drive him back, but Asif said he was first going to visit someone living nearby and preferred to walk. Not being very bright, Rafi took it for granted that he was going to see a mutual friend who lived nearby, Lt Col Naseer, in order to enrol him also. When he met Naseer next day he starting talking about Asif's visit and what he had said, only to discover that Asif had not in fact met Naseer. However, the cat was out of the bag and Naseer wanted to know what it was all about; he then told Rafi to report the matter, threatening that otherwise he would do so himself. Trapped, Tariq Rafi went to the authorities and was told to get actively involved and find out as much as he could.

Besides acting as an informer Tariq Rafi also played the role of agent provocateur. A couple of days after Muzaffar Hamdani had told me that army intelligence was aware of these activities,

I attended a meeting of some of the principals (including Tariq Rafi) at Asif Shafi's house in Lahore. I told them about this and strongly urged them to at once cease all activity. They agreed to do so, but after I had left Tariq Rafi persuaded them to continue with their plans without me (apparently Aleem Afridi and Farouk Adam didn't need much persuading). He even offered to arrange for a safe house where a meeting of all the persons involved could be held soon in order to make the final arrangements. It seems that before this could happen the army decided to act and arrest us all.

Chance also played a big role in causing my old friend, Ateeq Ahmed, to end up in Attock with us. I had talked to him once about the matter but that was about all. He was commanding the divisional artillery in 18 Div and had set out on a reconnaissance trip to familiarise himself with the large desert terrain along the border with India for which the division was responsible. Driving along in the empty wilderness on a desert track he came across another beflagged jeep which turned out to be carrying a friend of his – Brig Zaheer Alam Khan. During their brief chat Ateeq mentioned that he had heard talk about some officers thinking of removing the government. Zaheer did not show much interest, and they soon went their respective ways.

Attock Fort

Farouk Adam Khan

Asif Shafi

In front of my cell

Standing: (*L to R*) Iftikhar Adam Khan, Nasrullah Khan, Muzaffar Hamdani, Iftikhar Ahmad.

Sitting: (*L to R*) Ishtiaq Asif, Sajjad Akbar (*"Catlow"*)

When the 'plotters' were arrested, one of them happened to be ZA Khan's brother, an air force officer. Apparently, ZA Khan then struck a deal to save his brother by betraying Ateeq and later testifying against him at the court-martial. (While ZA Khan's brother was let off, he himself was retired from the army soon after, as were several others who appeared as witnesses against us).

In spite of the High Court's directions that we be treated according to army regulations, our conditions of imprisonment were still not meeting that standard. On 2nd June I wrote an application to the High Court bench saying their orders were not being complied with. This was held up somewhere in the army chain and was not placed before the judges until the morning of the 22nd, when they resumed hearing of the writ petition and I appeared again before them on their summons. They were naturally upset about this and ordered the Advocate General to file a contempt of court charge against whoever was responsible for delaying the application. In my statement I apprised them of the current conditions at Attock. Aleem appeared before them on the 25th and also gave his statement. On the 26th the bench directed the Advocate General to ask the army chief to take appropriate disciplinary action against the officers in-charge of Attock Fort for disregarding their earlier order, and to communicate the results to them by July 10th.

Obviously fed up with this meddling by the courts, Bhutto soon after issued a presidential order amending the Interim Constitution so that civil courts no longer had jurisdiction in respect of persons subject to military law. When the High Court bench reconvened on July 10, the AG argued that they could no longer proceed in the matter. The bench instead referred the matter to the Chief Justice to convene a larger bench to adjudicate on the constitutional issues involved. Meanwhile, the government had a Supreme Court judge suspend the High Court's earlier order requiring the army chief to take disciplinary action against the Attock Fort commanders.

As the Summary of Evidence progressed we began to push back against the restrictions imposed upon us and our conditions of imprisonment. Our jailer, Lt Col Aziz, resisted as best he could, but there were obvious limits to how much he could do to us – after all, we had to be presented in fairly good shape before the impending court-martial. Gradually we stretched out the time we were out of our cells, and thus able to congregate, until the only time we were locked up was at night. During the day we practically had the run of the place, and the guards would keep out of the way and mind their own business.

One day a group of us was sitting and chatting in a large barrack room that had been set up with seating arrangements for family visits when Aziz came in escorting a general officer. I recognized him as Maj Gen Anwar Qureshi, the Adjutant General, a nasty piece of goods; he was apparently there on his first inspection visit. He was quite taken aback to see us sitting there, relaxed and chatting, and told Aziz to send us back to our cells. As we left, Nadir Parvez and a couple of others began to chant, *Joy Bangla*! (They had seen during the recent fighting in East Pakistan how Mukti Bahini fighters would often chant this cry of defiance when lined up to be shot! This had had such an effect on them that this cry became for them the ultimate symbol of defiance).

After the Summary of Evidence finished they allowed our families to visit. Jamila (accompanied by various relations) came thrice, the last time with her brother Anis, who had recently been repatriated on medical grounds from an Indian prisoner-of-war camp (he had got caught up in the East Pakistan debacle). Many of the families brought children, and the more adventurous among them would play around in the open area and explore the cell blocks in which their fathers were living. One of these was my 7-year old nephew Omar (son of Iftikhar, Jamila's brother). He discovered Nasrullah and Munir Rafiq playing chess in front of their cells, and volunteered that he could also play. So Nasrullah played against him next – and lost. Thereafter, every time Omar came to the Fort he'd make a beeline for Nasrullah's cellblock and they would play chess, with several others watching. As often as not, Omar would beat him.

8

The Trial

(9 July 1973 to 1 Jan 1974)

There was much interest in, and speculation about, who would preside over our trial. Finally, one day Maj Gen Zia-ul-Haq turned up at the Fort to inspect the facilities, and it became known that he was going to be the one. He met us, expressed concern and annoyance at the conditions we were living in, and gave orders for a small officers' mess to be set up so we could get proper food and be able to eat it in a suitable place. He also arranged for a telephone to be made available so we could occasionally call our families. These measures were naturally welcomed, and led some to become a little optimistic about the trial.

It began in early July. The General Court-Martial was headed by Zia and comprised two brigadiers (Jahan Dad Khan and RAS Bokhari), a colonel (from AG's Branch in GHQ), two lieut. colonels and a major. There was also a Judge Advocate as the court's legal adviser. The prosecution was initially conducted by a retired military legal officer, but he was soon replaced by a Deputy Attorney-General of Pakistan. The accused comprised 19 serving and two retired officers (Aleem

Afridi and me). The former group consisted of two brigadiers (Ateeq Ahmed and Wajid Ali Shah), four lieut. colonels (Iftikhar Ahmad, Muzaffar Hamdani, Afzal Mirza and Muhammad Azam), 12 majors (Farouk Adam Khan and his brother Iftikhar, Saeed Akhtar Malik, Asif Shafi, Nadir Parvez, Ishtiaq Asif, Farooq Nawaz Janjua, Munir Rafiq, Sajjad Akbar, Tariq Parvez, Ayaz Ahmed Sipra, and Nasrullah Khan) and one captain (Sarwar Mahmood Azhar).

We were defended by a number of civilian lawyers, including two of the top ones in Pakistan: Manzur Qadir and SM Zafar. The former defended Aleem Afridi, the Adam brothers and me, and was assisted by Ejaz Hussain Batalvi, Waseem Sajjad and Aitzaz Ahsan (the latter two, young lawyers then, later became legal luminaries). SM Zafar, and his assistants, defended another group of the accused, while a few other lawyers represented the rest. Manzur Qadir was undoubtedly the pre-eminent lawyer in Pakistan; he had served as Chief Justice of the Lahore High Court for a couple of years, and also as Pakistan's Foreign Mininster. He defended us without charging any fee (we could barely scrape up enough money to cover his expenses), and in spite of his health not being good. He (and his assistants) spent long periods living in a rest house in Attock under fairly spartan conditions. Once he fell ill and had to be evacuated by helicopter to a hospital, but came back afterwards to resume his work for us. We can never repay the debt of gratitude we owe him.

The trial took place in a large converted dining hall in the Fort. On the first day the members of the court, resplendent in dress uniforms with clanking medals, mounted the dais where they were to sit. But they were not a patch in talent or valour to those they were to try. Dressed in plain uniforms, these sat at tables along the two sides of the room with their counsel, several of them wearing simple ribbons denoting the gallantry awards they had won in battle (Aleem Afridi, Farouk Adam, Nadir Parvez, Farooq Nawaz). Many of them had been platoon commanders at the Military Academy, positions filled only from the pick of the army. At the other end of the hall, facing the court, were seats for family members of the accused and other permitted visitors.

Jamila and Shamim Afridi, who were close friends, attended almost every day of the six month long proceedings. For part of this time they even lived under quite primitive conditions in an old house in Attock village. They also looked after the food and other living arrangements at the rest house for Manzur Qadir (and, often, his wife) and the other lawyers who lived there. Jamila's brother, Ijaz, helped a lot by regularly sending supplies from Rawalpindi. From time to time, Afroze, my brother Daniyal, and other close relations also visited and attended the trial, as did relatives of several other accused.

Gen Zia (known in the army as 'Terry Thomas', because of the resemblance) made a deliberate effort to show that the trial was being conducted in a fair and impartial manner. This was partly due to the presence of the senior civilian lawyers, and partly to his desire to avoid the sort of rowdy non-cooperation that had once occurred during the Summary of Evidence when the recording officer showed blatant partiality. Of course, sometimes the mask slipped: one Monday morning he started the proceedings by announcing that he'd had a very useful meeting over the weekend with the Director of Military Intelligence which had "cleared up many gray areas of the case" for him. Manzur Qadir wasn't present that day but when he raised the matter at his next appearance, Zia, in typical fashion, flatly denied he had said anything of the sort!

The proceedings were conducted according to military law and procedure; witnesses gave their evidence and were cross-examined, etc. Of course it was all a charade, we knew what the outcome would be, though some of the accused harboured hopes that the minimal evidence against them might lead to their acquittal. Tariq Parvez (a cousin of Nadir Parvez, which was the sole reason he was roped in) formally asked to be acquitted at the conclusion of the prosecution case (as was provided for in the rules) since he had not been mentioned even once in their entire evidence, but his plea was rejected.

The prosecution case was primarily based on the testimony of Tariq Rafi, who had joined the "conspiracy" and met many of the accused (all the while trying to push it forward). Others were implicated through confessional statements the interrogators had obtained from a few of the accused through the usual trickery ("*There's very little against you; tell us what you know and we won't charge you*") They fell for it because their involvement really did amount to no more than some indiscreet talk, or even just being present during some such talk. A few remaining holes in the case were plugged by a few perjured witnesses who tried to save their own necks by providing the needed false testimony (such as Major Shahid Ataullah, who provided the key testimony the prosecution needed to make me subject to trial under military law). Manzur Qadir had no illusions about the outcome of the trial; as he told us, all he wanted to achieve was to properly prepare the grounds for his appeal in the Supreme Court. I was kept quite busy analyzing the statements of prosecution witnesses and preparing copious notes for Manzur Qadir to use in his cross-examination.

At the end of the prosecution evidence each accused could make a statement before the defence evidence was led. The principal theme repeated in these statements was the deep impact the 1971 war and its disastrous ending had on them, their shame, and the anger they felt at their leaders who were responsible. The most stirring narration was that of Saeed Akhtar Malik, who said in part:

> "When war became imminent, I took leave from the PMA and joined my unit, thanks to my CO who requisitioned my services. The next day the war started. But instead of glory, I found only disillusionment. With each successive day the truth that dawned on me became harder and harder to swallow. The truth was that we were a defeated army even before a shot was fired. This was a very bitter truth. With each corpse that I saw, my revulsion increased for the men who had signed the death warrants of so many very fine men. Yes, fine men, but poor soldiers, who were never given the chance to fight back, because they were not trained to fight back. When they should have been training for war, they were performing the role of labourers, farmers or herdsmen, anything but the role of soldiers. This was not '*shahadat*', this was cold-blooded murder. Who was responsible for this? I was responsible! But more than me someone else was responsible. People who got paid more than me were responsible. You and your generation were responsible. The distinction of making all the dubious records in history falls on us. This time we have established an unbreakable record. We have a generation which created a homeland. Then it tore and destroyed that homeland. It now continues in its efforts to blow into smithereens what is left of it. What about these men? What were some of these men, these callous, inhuman degenerates, doing when their only job was to prepare this Army for war? Were these men not grabbing lands and building houses? Did it not appear in

foreign magazines that some of them were pimping for their bloated grandmaster? Yes, Generals, wearing that uniform *(he pointed at Gen Zia-ul-Haq)* pimping and whoremongering!"

Another frequent theme was the admiration and affection they felt for their fellow accused, and the bond that bound them together. Farouk Adam Khan summed it up well when, in his statement, he quoted Shakespeare's Henry V at Agincourt:

"We few, we happy few, we band of brothers".

The trial ended on New Year's Day 1974. The previous day the court had deliberated *in camera* on the guilt or innocence of the accused – for just a little over an hour (as I observed from my cell). Their findings weren't announced, but based on the subsequent procedure we deduced that except for Tariq Parvez (against whom there was no evidence at all, his name was never even mentioned) we had all been found guilty. This caused some despondency among the few who had hoped the weak cases against them would lead to an acquittal.

Next day, New Year's Day, the court held its final session. Zia opened the proceedings with a little speech in which he thanked us for our help in enabling the trial to proceed smoothly, and mentioned the families that had so assiduously attended the proceedings. He concluded by wishing us all a Happy New Year and long and prosperous lives! He then asked each accused if they wished to make a 'statement in mitigation' before the court decided on the sentences (as required under the procedures).

The previous day I had proposed, and the others had agreed, that instead of everyone making pro forma statements I would be the only one to speak. I wanted to make an emotional plea on behalf of the young officers amongst us in the hope it would have some impact on the court, particularly its younger members, when they decided on our sentences. This is what I said to the court:

> "Some time ago, while reading reports of British court cases, I was struck by the fact that when the judge asked the accused whether he had anything to say before being sentenced, the reply sometimes was that the accused wished several previous offences of a similar kind committed by him to be taken into consideration. I am still puzzled as to how this could be a mitigating factor, but there must be something to it. So, when the Court asks me, "Do you wish to say anything in mitigation?" my answer is, "Yes. I would request this Court, before it sentences me, to take into consideration several previous offences of a similar nature that I have committed."

The first offence that I committed was that I, then Major FB Ali, on the 27th October 1958, together with General Ayub Khan and other general and field officers of the Pakistan Army, did wage war against Pakistan, in that they, using the troops under their command, did abrogate the legally enacted Constitution of the Islamic Republic of Pakistan of 1956, and did overthrow the legally constituted government of Prime Minister Firoze Khan Noon, and did seize power for themselves by means of criminal force.

As a young major I, like the bulk of the other officers of the Army, was not privy to the conspiracy that preceded this act of waging war, nor, like them, did I participate in this crime. But, morally, I am equally guilty for I was part of the Army before this action and remained in it afterwards.

My second crime was that I, then Lt. Col FB Ali, on the 25th March 1969, together with General Yahya Khan and other general and field officers of the Pakistan Army, did wage war against Pakistan, in that they, using the troops under their command, did abrogate the legally enacted constitution of the Republic of Pakistan of 1962, and did seize power for themselves by means of criminal force.

My third offence was that I, then Colonel F.B. Ali, on the 25 th March 1971, together with General Yahya Khan and other general and field officers of the Pakistan Army, did wage war against the people of Pakistan, in that they did abrogate the Legal Framework Order, 1970, and did take into custody the legally elected representatives of the majority of the people of Pakistan, and did use criminal force against this majority.

Then there were some conspiracies to which I was not a party. During December 1971 I found myself being involved in a strange conspiracy, not a conspiracy to wage war against Pakistan, like the other ones, but a conspiracy not to wage war against India. By then I had learned the importance of agreement in a conspiracy, and I made sure that my disagreement with this conspiracy was known to all. I even put it in writing.

On the 18th and 19th of December 1971, I became aware of another conspiracy in which I could become involved. This was being hatched by the junta in power to wage war against the people of West Pakistan who were demanding their ouster. This time, too, I made it clear that I, and the troops under my command, would not participate in any such mad venture.

I am well aware of the difficult situation in which this Court finds itself. The powers that be, and the compulsions of the situation, demand sacrificial victims. Very well. I offer myself to you for this purpose. I give myself willingly. No blame shall attach to you. Use me to propitiate the idols that must be propitiated. Do what you will with me. But let these others go.

You have held that a crime has been committed. But in your hearts you know that these are no criminals. They have all fought for this country. Many of them wear on their breasts the marks of honour of this country. Some of them have shed their blood for this country. All of them have risked their lives for this country. These are no criminals. Let them go. An army does not have many such officers. A country has few such sons. This army needs them. This country needs them. Let them go.

You have held that a crime has been committed. But what, really, is this crime? You know, everyone knows, that all this crime was talk. Angry talk. You say, dangerous talk, criminal talk. Very well, but talk still. And in your hearts, you should know that it would have remained talk. If you do not know, then I tell you that is so. For this talk they have paid the price. Their careers have been shattered. Their lives have been blighted. For almost one year they have lived the life of prisoners in jail. Indignity and humiliation have been heaped upon them. They have paid the price. Their families have paid the price. Wives have been widowed. Children have been orphaned. Parents have been bereaved. Whatever be the crime, they have paid sufficient price. Let them go.

You have said they are guilty. But they are guilty of many more things. They are guilty of having loved this country too much. They are guilty of having cared for this Army too much. They are guilty of having felt the disgrace and humiliation of this Army as their own disgrace and humiliation. They are guilty of having felt the wounds inflicted on this country as wounds inflicted on their own bodies. They are guilty of having cared too much and loved too deeply. It is because they are guilty of all this that I say to you – let them go.

Take me. Do what you will with me. But let them go".

At the end of my statement there was hardly a dry eye among my fellow accused and the spectators. Even the three younger officers on the court were obviously moved and had to conceal their emotion behind the hands on their brows. At its conclusion, Zia made a quite unnecessary and graceless remark: *Very moving, but the law and discipline are blind.* Still, I think it did have some effect; we learnt later that the three younger members of the court-martial had tried to limit the sentences imposed.

After my statement the court went into closed session to decide on the sentences and we were sent back to our cells. Mine overlooked the building where the court was meeting and I kept watch to see how long they would take. They had decided on our guilt in about an hour, but they took hours to decide on the punishments. At one point the youngest member of the court, Major Muzaffar Hussain Usmani[7], walked out, apparently in protest. Soon after, Brig RAS Bokhari came out and appeared to be trying to persuade him to return; it took quite a while before he succeeded. Finally, they came out, and I knew our fates had been decided (though we wouldn't find out what they were until

---

[7] Usmani later earned a mention in Pakistan's history when, in 1999, he was a Lt Gen commanding the corps in Karachi and took over Karachi airport to allow Gen Musharraf's plane to land, thus aborting Nawaz Sharif's attempt to replace him.

much later when they were promulgated after the army chief's approval).

* * * * *

On the night of 4th January, Aleem and I were suddenly and without warning taken from the Fort to a small army barracks near the jail in Campbellpur; it had been prepared as a prison with a barbed wire perimeter and sentries. No explanation was provided for this move, which seemed particularly odd considering the next day was Eid. It seemed to me the only reason could be that we had both been sentenced to death and had been brought here so that, as soon as the sentences were approved by the army chief, they could be carried out quickly and without fuss. However, when nothing had happened after several tension filled days, I relaxed a bit.

We later found out that my fears had not been misplaced. When we later went to Campbellpur jail we learnt the officials there had been ordered at that time to prepare the gallows for executions. After I came out of prison I discovered that the court had sentenced us both to life imprisonment but Bhutto expressed his displeasure and the army chief, Tikka Khan, sent the papers back to the court for reconsideration, with a directive recommending the death penalty as a more appropriate sentence. Court-martial sentences are decided by the votes of members; to change ours to a death penalty five votes (out of the seven) were needed but the three younger members voted against it. It was a close call.

Aleem and I stayed in this *ad hoc* prison in Campbellpur for about seven weeks. Finally, on March 1st we were told that promulgation of the sentences would be carried out the next day at Attock Fort. On the morning of the 2nd we were taken back to the Fort – and a joyous reunion with our comrades. We found them locked up in their cells, and soon we were also locked up in our old cells. At about 10 a.m. I was handcuffed and taken under guard to an open space where Brig Ghouse stood at a lectern within a hollow square formed by a ragtag collection of soldiers from the Fort garrison. I was positioned in front of him and he read out the charges against me and then announced the verdict and the sentence of the court: imprisonment for life. I was then taken, still handcuffed, to a truck parked nearby.

Gradually, one by one, the other 15 sentenced to prison joined me. Aleem also got life, Farooq Adam 10 years, Nadir Parvez and Asif Shafi 7 years, with the others getting lesser terms, most of them 5 years. Of the remaining five, a couple (Brig Ateeq Ahmed and Jamila's brother, Lt Col Iftikhar Ahmad) were dismissed from the service while the rest were given other lesser punishments and then retired. The sentences were vicious and brutal, especially in the case of the several young officers going to prison who had only been proven guilty of some indiscreet conversations. I suppose Bhutto wanted to send a clear and stark message to the military.

The truck drove off towards Campbellpur jail, a dozen or so miles away, with us in the back. Jamila and Shamim were parked outside the Fort gate in the latter's car, and began to follow the truck. They gestured at us wanting to know what our sentences were; we tried to respond but it is difficult to indicate a sentence of imprisonment for life using ten fingers of cuffed hands.

* * * * *

A long, hard road had we traversed, with no end yet in sight. This catastrophe had drastically changed so many lives and shattered others; many of them so young. The babies that were born to fathers in captivity (Asif, Munir, Farooq Nawaz, Nadir Parvez). The murdered brother (Munir's). The father who died of a broken heart (the Adams'). The many others who walked about and smiled and even laughed, but within whom something had died, and something more was dying as the lost weeks and months went by.

It was heart-breaking to see the children, and difficult to decide who was worse off: the small ones who didn't understand what was happening and why their fathers would not come and be with them; or those old enough to understand only too well and perhaps also peer into a frightening future. And feel the hurt at what was being done to someone they loved.

The time of Attock was a strange time. It started with chains and blindfolds and iron bars. Then there were times when it became almost a holiday camp. Interspersed throughout were the crises and confrontations, and then there was the fraud and farce of the trial. The grinning Terry Thomas who played the role of the kind and fair judge right up to the last day when he drove in the final nails into our coffins while wishing us all (even the wives) a Happy New Year and long and prosperous lives!

Cut off from the rest of the world, Attock was a world on its own. With its resident snake (Aziz) and a visiting clown (Ghouse). A world we never made. Yet we lived what seemed our whole lives in it. Its name will be found branded on our hearts (how many more such names will keep it company?). We lived there, we laughed our laughs, we shed our tears, and then we left it as we had come to it – in chains.

9

Campbellpur

(March to July 1974)

It was about an hour's drive from Attock Fort to the prison in the nearby town of Campbellpur. The sentences announced at that stupid promulgation ceremony had been brutal. Some of the others had not expected to be imprisoned based on the evidence of their involvement presented at the trial, and it had come as a big blow to them. For all of us there was the uncertainty and nervousness attending our passage from military custody (a familiar environment, however bleak, and one in which, even as prisoners, we had some standing) to a civil prison, of which we knew nothing and where we were just prisoners like any others.

Our apprehension was not allayed when we learnt that the army had given us 'C class' status. This classification system was a hangover from the days of the British Raj and determined the facilities an inmate was given in prison; C was the lowest category. Accordingly, we were taken to a couple of bare rooms without any furniture in the small prison hospital and were each given a mat and a couple of filthy, tattered blankets to put under the bedding we had brought with us. We were told that since the rations for the day had already been issued, there were no meals for us that day. Luckily, our companions who had not been sent to prison soon arrived to see us, and they arranged for our food for this first day. It was a depressing beginning to the long journey that lay ahead.

Over the next few days there were several false reports that our class had been upgraded. These were caused by some strange goings-on in the army. Messages came to the jail superintendent to this effect and were then cancelled. Once Brig Ghouse even came there to sign the necessary papers, then telephoned GHQ and left without signing. There was even a call from the army chief to the superintendent enquiring about the facilities available under each class (typical of Tikka Khan to find out himself)! Finally, after five days, on 7 March, we were given 'B class', which meant we didn't have to sleep on the floor any more, and could use some of the better rations now provided instead of buying everything from the prison contractor. It took until May 10, and a visit to Bhutto by Munir Rafiq's in-laws, influential party supporters in Sind, before our class was at last upgraded to A.

We had been allocated the hospital kitchen and provided with a couple of prisoners as cooks. Asif Shafi took charge of this operation and pretty soon we were eating quite reasonable food. Since we were all sentenced to "rigorous imprisonment" we were expected to work in prison. A couple of days after our arrival four of us were assigned to assist in the offices while the rest were detailed to work in the carpet factory; they were quick to add that this would only be on paper, though we should report there at the start time and then we could return to our cells. However, the Superintendent wasn't happy with this arrangement made by his subordinates, and we had a bit of a tussle with him over the next few days. He finally gave up and told us that so long as we remained in the factory during work hours, we could do what we liked there. This suited us fine since the factory was a large open space (it also housed the gallows), and we preferred to spend the mornings there reading, playing cards, or chatting, instead of in our smallish dorm cells.

After a few days our life settled down into a fairly regular routine. As I noted in my journal:

> "The warder unlocks our cell block at around 6.30, switches on the lights and does a headcount (literally so, as all he sees is our heads on our pillows). Some of us get up quite early (Azam and Wajid get up for prayers). I get up at about 7 or 7.30 and start getting ready (usually I am the first to do so in my room). Breakfast is laid out on the dining table in the other room – boiled eggs (though some days there's an omelette), toasts, butter, jam and tea. Most people have breakfast between 7.30 and 8.30, wandering in at different times.
>
> Between 8.30 and 9 people start moving in ones and twos to the factory. Chairs are carried from the rooms and placed under the canopies on the two (disused) waterwheels there. We sit around reading or chatting till about 10 or 10.30 when the tea comes. Before this the Superintendent or his Deputy comes on his rounds, and usually stops for a few minutes to chat. After tea most people wander back to the rooms, usually to start a game of cards or chess. I like to stay on; it's a change, and besides in the factory one doesn't have that closed-in feeling.
>
> Lunch is about 1 or 1.30, after which people generally lie down and sleep or read. At about 4 or 4.30 the volleyball net is put up in the yard in front of our rooms, and the game starts. About 7 or 8 of us play along with quite a few other prisoners, while a large number of others sit around watching. The games arouse quite a lot of interest in the prison.

I start my exercise at about 5.30. Half an hour of rapid walking followed by other exercises. The volleyball players finish about 5.45 and start going one after the other for their baths. At 6.30 we are locked in: the door to the outside in the barred verandah is locked (but we can still move between the two rooms we have). In the evening people read, play cards or chess, and chat. Dinner has been cooked by 6 and the pots are brought in. It is reheated and eaten at around 8, then more reading and chatting. We usually turn in at 10.30 or so.

There's a fan in our room. We use it in the daytime, but not yet at night. Farouk Adam wants it on but gets outvoted. We have a democratic system of deciding these things. For lights, when 4 out of 6 decide to go to sleep the lights go out. Usually Farouk wants to continue reading and complains loudly.

Most of us wash our own clothes. There is a *dhobi* (washerman), but he didn't wash my clothes very well the first time I sent them to him. So after that I started washing them myself and just get them ironed by him".

The jail was a large area enclosed within a high wall. Spaced around inside were the various wards, with walls enclosing their cell barracks. There was a high tower in the centre of the prison called the *chakkar* from which the guards on duty could observe the whole prison. Another fairly large walled off area housed the "factory", which contained the carpet-weaving operation that the prisoners worked in. During the day the prisoners were able to move around fairly freely in the quite substantial open spaces in the jail, and one didn't really feel confined.

Campbellpur Jail

In the verandah of our cell block

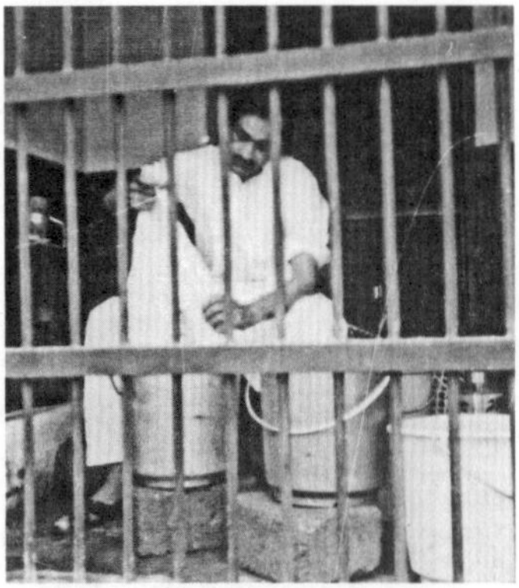

Washing my clothes

Tea break in the prison 'factory': (*L to R*) Iftikhar Adam Khan, Nadir Parvez, Me, Aleem Afridi, Wajid Ali Shah, Munir Rafiq, Farouk Adam Khan, Saeed Akhtar Malik, Ayaz Ahmad Sipra.

One evening early in May the Deputy Superintendent took us up to the top of the central tower. This was the first time in two months that we had seen the world outside the prison walls. It was lovely. On one side the golden fields with their green borders stretching into the distance, and on the horizon the dark hills of Attock. On the other side, houses, people, a world of strange normality. There, close by, was the hut in which Aleem and I had lived for a couple of months before coming here. After drinking their fill some said: *I wish I hadn't come up here, now I will feel bad.* From the tower we looked down into Death Row and saw the inmates, each locked in his cell, standing at their barred cell doors looking up at us. To them we must have seemed very lucky.

There were the typical sounds of jails that we were so aware of in the beginning, but gradually began to notice less and less as time passed. A redcap is always on duty at the top of the central tower and throughout the night he is supposed to call out to the redcaps responsible for the various wards, and they are supposed to shout back that all is well. (However, the callers are not very punctilious about the calling, and often the lower redcaps don't answer back). There is also a bell in the central tower which tolls the hours – or some of them at least – twenty minutes early. Two bells are rung to announce distribution of food, once in the morning, once in the evening. A single bell announces the entry or exit of some official from the jail. The bell is also supposed to sound the alarm (or "Larrum", as they called it there) by continuous ringing. (One of the earliest pieces of advice given to us was that if we heard the "Larrum" we should quickly go into our rooms and stay there because, when the alarm is sounded, the warders usually run wild beating up anyone found outside).

When we arrived at the jail we were objects of great curiosity among the inmates, and throughout the first day a constant stream of them passed by our area to gawk at us, with the bolder ones coming in and talking to us. Among these latter were quite a few former soldiers; one of them had actually served under me in my regiment and appeared genuinely distressed to see me there. One of the prisoners who seemed to have some standing became quite friendly with some of us and tried to help with advice and guidance. His was an interesting story: Daud Khan was a Hur from Sind who had remained a 'proclaimed outlaw' for 7 years, during which time he claimed to have killed many people and even kidnapped the Deputy Commissioner of a district. When finally caught, he had been sentenced to 112 years in prison (of which he had completed 12). Quite a character.

During our stay there we got to learn quite a bit more about the prison and its inmates. As I wrote in my journal:

> "Jails are funny places. A world of their own, cut off almost completely from the real world. Prisoners are just about the lowest of the low; they have no rights and are completely at the mercy of the jail authorities. It seems that as a matter of policy every new prisoner gets slapped around on arrival and for the first few days. To get him in the right frame of mind, I suppose.
>
> The inmates are also a strange lot. There are a couple of old parties who are over 80 (or so they claim); they do look pretty old, but appear quite fit. There is a father and son (probably more). Two brothers, one serving life, the other on death row (the former was an FC College student). There is a husband and wife, and also a brother and sister. There are said to be four women in the Women's Ward. I once heard a baby cry in this ward and learnt that this woman was two months' pregnant when she came here and the child was

born in prison; she is still under trial for murder. There is also a Boys Ward. Two boys, 6 and 8 years old, came here on a murder charge; one of them may still be here. There are about 25 people on death row whose cases are at various stages of appeal; some of them have exhausted all appeals and are now awaiting decisions on their mercy petitions. However, they say there have been no executions since Bhutto took over. There are two people sentenced to death whom everyone says are innocent.........

I have written earlier about the poor lot of prisoners. But looked at from another angle, life is not really too bad. There's very little work that they do. For example, in the factory they start work at about 7.30, but after the Superintendent's round at about 9.30 I see very few people working; most of them either lie down and have a nap, or sit in groups chatting. They remain thus till about 3.30 or 4 when they return to their barracks. Others not working in the factory also seem to have a similarly easy time. Once a person gets reconciled to the loss of freedom, life would not be too bad. Sufficient food, which probably compares well with what most of them eat at home, leisure, company, in an odd way security, no problems – all these are plus points. Perhaps that is why most of the people here don't seem to be upset at being in jail, and appear quite cheerful........... .

Daud, the Sindhi outlaw, told Munir one day about executions that he had seen in other jails. The night before, a large crowd of people comprising the condemned man's relations and people from his village arrives. They spend the whole night outside his cell, singing and dancing. Usually, Daud says, the condemned man keeps shouting religious slogans all night, and by morning his throat is hoarse. All sorts of good food is brought for him, but he rarely eats anything. Then at 4.30 in the morning he is led from his cell to the gallows. Many of them stamp their feet and shout slogans as they proceed and other prisoners join in the slogans from their cells. At the gallows the condemned man is asked his last wishes, if any. Then the deed is done, and his body is handed over to the waiting crowd of relations and friends".

After the initial shock of landing in this prison, and the first few days adjusting to our new life here, most of us were in fairly good spirits. This was mainly due to our expectations regarding the plans our lawyers had made to approach the high court to overturn our convictions or sentences. We kept getting news about the various discussions and steps that were being undertaken, and our spirits rose and fell accordingly. Finally, a writ petition was filed in the Lahore High Court, but before it could be taken up Bhutto had an amendment to the constitution moved in the National Assembly ousting the jurisdiction of the civil courts in the cases of serving and former officers of the armed forces. Jamila and Shamim, with the assistance of some of our supporters, tried to create a public fuss about this. I drafted a note on the amendment which they slipped under the door of every room in the hostel where the National Assembly members stayed. Their efforts had some effect because the reference to "former officers" was removed from the amendment before it was passed by the Assembly.

That didn't prove of much use, because the Lahore High Court refused to hear our writ petition and dismissed it; it also refused to issue a certificate to allow the lawyers to lodge an appeal in the Supreme Court. So the only recourse left was to initiate a fresh case in the Supreme Court, and Manzur Qadir did not want to do this before October. In any case our chances now hung on this one gamble. The emotional roller coaster that we rode during this period was also caused by other news that kept trickling in. Many people were trying to use their influence on our behalf, some with the military, others with politicians. Sometimes hope would be kindled, only to be soon extinguished. We finally realized that anything to do with us could only be decided by Bhutto, and he was in no mood to show us any leniency. He even went back on his promise to Munir Rafiq's in-laws, who were influential supporters of his political party, to reduce his sentence.

One of the episodes that raised our spirits for a while was the news in April that Bhutto had summarily retired the Air Force chief, Zafar Choudhry, and several other officers involved in the investigation and prosecution of the Air Force officers tried in the PAF court martial that ran parallel to ours. Seven of the convicted officers thrown out of the service were also reinstated. The PAF had roped in a large number of officers, many well liked and respected in the service, often on flimsy grounds with several suspected to be victims of a settling of old scores. The unhappiness this engendered, reinforced by reports of the brutal treatment meted out to them in captivity, became acute when it was learned that the Air chief had leaned on the court to ensure convictions and heavy sentences. Soon after the sentences were announced about 70 to 75 officers, including many pilots, submitted their resignations. Alarmed by this, Bhutto acted swiftly to distance himself from this affair. Many of us hoped he might do something similar in our case, but nothing came of it.

It was not only the loss of our freedom and the blighting of our careers that weighed on our spirits but also the separation from our families and the impact of our imprisonment on them. I wrote about this in my journal:

"This is one aspect which hurts me greatly, even though I do not let myself think about it often. That is the shattering of so many families, the widowed wives and orphaned children. Perhaps it is harder for those with younger children; it is these years of their lives that one least wants to miss. It hurts. This whole thing hurts. On her last visit Afroze told me that Asif Shafi's sister had said to her about his wife: "*Sara din veeraan phirti rehti hai. Aur issy tara bachhe*" ("She wanders about all day, lost and desolate. And it is the same with the children"). It was like a stab to the heart.

Except for Azam no one talks about this aspect. But in each one's heart this cancer must be eating away. Each day that passes is a day lost, a day less in the time one has to spend with one's wife and children, a day that can never come back. They never talk about it, but sometimes one can glimpse the rage, the hurt, the sadness in their hearts".

Still, one of the things that sustained us in this difficult passage was visits by family. For me especially, the fairly frequent visits by Jamila were invaluable. I could imagine how difficult it must be for her to make the travel arrangements, based as she was in Lahore with my mother. But she and her friend Shamim Afridi were indomitable, refusing to accept the hand that Fate had dealt us, pursuing the lawyers on our cases, working on politicians and journalists on our behalf, looking after the children, keeping up our spirits in prison. They were both superb!

Not everyone was so fortunate; many families could not come often to this remote place. But family visits also had their dark side. They were wonderful while they lasted, but afterwards the separation was that much harder to bear. This was captured in an entry in my journal for June 8:

"Nadir's brother came to visit him today and brought along Nadir's two children – a little girl and a younger boy. When they were leaving, the young boy kept asking Nadir to come along with him (how do you explain to a boy so small that you are not staying away of your own will?); the girl, only a little older, at least knew that "Bhutto had locked up her father". Nadir was visibly upset the rest of the day. Later on he confessed that, while having a bath that evening, his pent-up feelings burst and he suddenly found himself flinging the mug in his hand at the wall and loudly cursing Tikka Khan, calling for him to find himself in the same position in which he had put Nadir. I can imagine the fervour with which he must have done this; I wouldn't be surprised if his tears were flowing at the time".

Finally, on the subject of family, my journal entry of May 18 perhaps captures in its few words the unspoken hurt and yearning that we all carried in our hearts: "*Jamila came to see me today. I have never seen anyone look so lovely in all my life*".

My journal has many entries recording the ups and downs of my spirits in these four months that we spent in Campbellpur jail:

"In the evening when the sun has set, and before it gets dark, walking about in the yard, if one looks up at the sky, for a while one can forget that one is in a prison and savour the illusion of freedom. The birds flying free to their night's rest. The clouds floating free where the wind carries them. The deep blue of the sky, untainted, unlimited. Sometimes a breeze, unhindered by these ugly walls, coming and going as it pleases. A peace and quiet that belies the ugliness of captivity. For a few moments one breathes deeply and feels free".............

"A sadness and a weariness overcame me. A wanting to be done with this constant struggle, these ups and downs, to just give up and accept the worst. Sadness at the waste. At the defeat. A fate undeserved? More doubt replacing a belief whose foundations were eroding. A belief that had seemed logical enough in the abstract, but could not account for a personal fate that seemed increasingly to be the plaything, not of blind random chance, but of an antagonistic purpose".........

"My reserves must be getting low: I was quite depressed by yesterday's news. Or is it that deep inside I hope too much? On the level of consciousness I am fairly calm and composed. In Attock and later here I have been pretty steady with few ups and downs. I have never worried about the ultimate outcome, having had no difficulty in maintaining the assumption that it would all turn out well. I have endured the conditions of imprisonment very well; I do not get depressed or on edge; most of the others have such phases".........

"I very seldom think of the future, or even dream or hope. Yet sometimes when a corner of the curtain lifts, I get a glimpse of the turmoil behind it: the anguish of a deep hurt, the fierceness of a strong desire; the hurt of the bird cut down in mid-flight, the desire of the broken wings to fly again, the anguish of the memory of the green glades. But on the surface, all is calm. I do not get depressed as most of the others do at one time or another. I am the one who tries to cheer them up. But underneath?"..........

"One can see how slowly, quietly, each in his own way, is sliding into a state of desperation. Not too deeply yet, but the slide is there"........

"Sometimes, sitting outside, I look at the sky, at the stars, and I ask myself: What am I doing here? There is a whole world, a whole universe, outside these walls; what am I doing here? Sometimes one catches one of the others looking around, and from the look in his eye one can guess that he is also asking himself the same question: What am I doing here?"..........

"Will we ever look back on this time, and be able to say: Well, it was worthwhile?"..........

"Does it all mean anything? Does anything mean anything?"

When we arrived in Campbellpur jail we knew that it was going to be a temporary stay, and that soon we would be sent to different jails. This prospect of our group being split up, with no knowing when we would be together again, or even see each other again, was a sad and daunting one. When we had arrived at Attock we were a disparate group, with many being strangers to each other, or mere acquaintances. What we went through there together had forged close bonds among us; we had truly become a band of brothers. Knowing that we would soon be scattered was a secret sadness and fear that each of us carried in his heart, and it made whatever remaining time we had together all the more precious.

Over the course of the next four years that I spent in solitary confinement in various prisons I treasured the memories I had of my companions and our life together, especially in Campbellpur. Asif Shafi, a true son of Lahore: solid, steadfast, large-hearted, ever ready to meet each new adversity with a quip or a choice Punjabi epithet. Practical and very good with his hands (one evening he and I spent quite some time excitedly making plans to start an electronic repair service in Gulberg 'afterwards'). He was generous, helpful and considerate, and I remember the many kindnesses he showed me, especially when I once fell ill and lay in bed with a high fever for several days.

Farouk Adam, a born leader, but also an incurable optimist, betting whoever would wager that we would be out by his birthday on May 24. Saeed Akhtar Malik, having had to grow up in a hurry in Attock, his beard and sober manner now belying his young years, whose wise counsel was always worth listening to. Nadir Parvez, larger than life, his bold spirit manifesting itself in the swagger of his walk and his booming voice. Wajid Ali Shah, walking in the yard in his immaculate white *shalwar kameez,* the personification of dignity and grace. Muzaffar Hussain Hamdani, graying, stocky, older than most of us, never talking about the fact that his family was probably the hardest hit of all. Young Sarwar Mahmood Azhar, the 'baby' of the group, for whom life was still a lark and prison a great adventure.

Munir Rafiq, that first night in Campbellpur as we lay in the dark on our makeshift beds on the floor, speaking with great feeling about the effect his going to prison would have on his family (he had got married a few months before his arrest, and his first child had been born while he was in Attock). Iftikhar Adam one day gravely explaining to me his theory of how evolution could be reconciled with religious belief. The evening when Nasrullah had just finished regaling us with a song (as he sometimes did) and a female voice rose up in another song from the women's ward nearby, and then they both sang alternately for quite a while (it turned out to be a 'singing girl' locked up there for a couple of days until she was bailed out).

We were also concerned about where each of us would be sent from Campbellpur. Under the rules prisoners serve their sentences in a jail close to their homes, but we couldn't be sure that would apply to us. Especially since it wasn't GHQ or the province that was controlling our fate but Bhutto himself; we had no idea what he would do with us or when he might finally wash his hands of us. The weeks stretched into months and, apart from rumours, nothing happened. Then, suddenly, on Saturday, July 13, we were told that we were all going to move, starting with some that day. The first lot comprised Asif Shafi and Saeed (going to Mianwali), Munir and Nasrullah (to Sahiwal), and Hamdani and Azam (to Lyallpur). These jails were not near their homes; it seemed we were being scattered to far-off places, ones that were also not near military cantonments.

I recorded our departure from Campbellpur jail in my journal thus:

> "Around 1.30 Farouk Adam came and told me about the imminent transfers. Soon rumours started that we were to be flung to far-off places. Farouk asked me to go and speak to the Deputy to find out some more, and also to ask him to provide us with due notice. The Deputy was very secretive, and would only say that some were going today and others on Monday, but wouldn't say where.
>
> I was already feeling low. I had been reading the Qur'an and had come to a verse whose implications set off a whole chain of thought which left me feeling shaken and forlorn. Despite my doubts as to whether I could expect anything from God, I suppose in my heart I must have had some hope. Otherwise, I would not have felt so desolate. Because, reading these verses and thinking out their implications, I could only conclude that God does not interfere in the affairs of mortals. My thoughts inexorably led me to the conclusion – there is no God on earth. And accompanying it was this awful feeling of being alone, lost, abandoned.

Soon after, the list of people going was sent to us (though not their destinations). People started packing, while bits of information trickled in. Asif Shafi was quite upset to learn that he was going to Mianwali. So was Saeed. Moves from jails always seem attended with panic and rush. Farouk Adam is a sentimental person, and he was quite affected by the occasion as he sat in our room while Asif packed.

Soon they were packed and we all assembled near the inner gate to say farewell. Quite a few other prisoners were also there to say goodbye. It was a sad occasion. We had been together for so long, we had gone through so much together. Now we were parting, and no one knew when we would meet again. It was doubtful if we would ever be all together again.

The final embraces. The drawn, worried•faces trying to smile. The sadness of an ending. Sipra embraced Asif Shafi and walked away with tears rolling down his face. Farouk's eyes were wet, so were many others. Then the door in the big inner gate opened, and they were gone. The remaining few of us returned to strangely empty rooms.

By that evening I had learnt that I was going to Bahawalpur or Bahawalnagar. I didn't even know where the latter was, but they were both very far away from Lahore. I accepted this, but I knew my family would be very upset. I had had a telegram sent to Jamila earlier about my leaving, but at that time I didn't know where I would be going.

I was resigned. When there is no God, then one is at the mercy of men. There was nothing that I could do except survive. That night I lay awake for a long while, a deep sadness in my heart, trying to comprehend a world without God.

I got up late the next morning. I had planned to slowly wind up things in preparation for the next day's move, but at about 9 a chit came that the rest of us (less Aleem and the Adam brothers, who were to be transferred to the Frontier province) were to get ready to leave at once. Since morning the electricity had been off, and there was no water in the tap.

I had just finished breakfast. I brushed my teeth and shaved, and then started packing. Reminders started coming asking us to hurry. I told them I would not move without having a bath, and they should arrange some water. Anyway, finally I was given a half-bucket and had a 'bath' with it, and then finished packing in a rush with the Deputy standing around wringing his hands and moaning that we would miss the train.

Once again the small group at the gate; the embraces, Farouk Adam in tears. And then the door opened, and we were through the gate".

10

Bahawalnagar

(July and August 1974)

This was a small, tattered town far away from the beaten track, stuck in the middle of a desert. The jail was very small, about a quarter the size of the one at Campbellpur. It was terribly overcrowded; they told me they had about 400 inmates even though the prison was meant to house only 130. They had no place to put me up, so they emptied a two-cell block off to one side of its 31 inmates. It was in terrible shape – plaster peeling off the walls, dirty broken floors, dirt and insects everywhere. They produced some prisoners to clean, patch up and whitewash the place, which took a number of days.

I had these two cells with a small walled off enclosure in front, though I used only one of the cells. In the enclosure were a couple of small, walled open-air cubicles, one serving as a latrine and the other as a bathing stall. There was a hand-pump in the yard which provided the water for drinking, cooking and bathing, although it tasted funny and left your skin feeling slimy after a bath. A prisoner was provided to do my cooking, initially on an open hearth made by stacking a few bricks, and later on an oil stove. His knowledge of cooking was minimal.

I soon encountered the twin scourges of B'nagar jail. By day it was flies. There were hundreds of them, constantly buzzing around, effortlessly dodging all attempts to swat them away, crawling over you with a few invariably falling into your plate of food. It took a while to get used to living with them. When evening fell, the flies gradually disappeared and the night scourge emerged – rats. Big, fat rats, obviously making a good living in the jail. The first evening as I sat in the yard I saw movement on the top of the enclosure's wall; there were a couple of rats there, enjoying the evening air. After a few minutes a couple turned up on the bars of the yard gate. I was not happy; rats were among my least favourite rodents.

Sure enough, that night when I went to bed, I soon heard them scurrying about the room. I shone my torch and tried to scare away those I saw. I heard a 'plop' and saw one had fallen into the bucket of water in a corner of the room; after swimming around for a while he managed to jump out. I knew it was only a matter of time before some of them climbed onto the bed. I kept awake for as long as I could, but must have dozed off when I felt something move against my arm; I sprang upright and the rat jumped off the bed. I again tried to keep awake, moving my arms or legs occasionally to discourage any further exploratory visits. I dozed off again after a while and awoke to find one of them nibbling on one of my toes. I didn't get much sleep that night.

Next day I asked the 'cook' to bring me a cat which he had told me wandered around in the jail. After some time he returned with a rather skinny looking creature that we fed and then secured with a longish string. That night I tied the other end of the string to one of my bedposts; I wasn't woken up by any rats but was by the cat, which had wound its string around the bedpost and started mewing, I unwound it and went back to sleep, only to repeat the exercise several times. When I got up next morning the 'cook' checked in and under the various items in the room and, sure enough, a rat jumped out of a basket but was promptly caught by the cat. After a few days of being chased whenever they appeared, the rats decided to stay clear of my cell, even though I'd regularly see them sitting on the yard wall and gate in the evenings.

The weather was hot and humid, and I felt exhausted much of the time. There was a ceiling fan but the electricity was frequently off. Adding to the stress of these unpleasant conditions were the onerous restrictions imposed upon me by the jail authorities. I couldn't tell if this was due to the fear they felt at having a 'high value' prisoner on their hands or just bloody-mindedness (it was probably a mixture of both), but they made life difficult for me. I had to surrender all the money I had; they also took away my writing material as well as all razor blades. It was only with considerable difficulty that I managed to get them to leave a notebook and pen with me, and even then they numbered all the pages. I was not allowed to step outside my compound and the gate was kept locked all the time. At night I was locked in my cell. I was used to exercising regularly, and missed not being able to do any walking.

Perhaps the worst part of this regime was the difficulties I faced keeping in touch with Jamila. I was told that every letter I wrote would count as a visit and reduce the allowed quota accordingly. Even to write one I had to send for my pad and an envelope, and these would take hours to arrive. After I had given them the letter to mail I had no guarantee it would be done (I learnt later that a couple of my letters had in fact been sent to their headquarters in Lahore for vetting before dispatch, as had some of hers to me). A couple of times Jamila phoned the Superintendent and gave him a message for me; they would take their own sweet time to deliver it; once not telling me until four or five days later. Being thus cut off was the hardest part to bear. The news I did get from her was in any case depressing: our lawyer, Manzur Qadir, who was to pursue our case in the Supreme Court, had fallen seriously ill and had to be taken to England.

The difficult living conditions and the aggravation of these petty restrictions left me quite depressed. The worst part was the frequent feeling of helplessness; I'd had no experience of this in my previous life. I sought to keep my spirits up so I could cope with all this – and all that was still to come. Parts of my journal entries of the period reflect my reactions:

> "The miserable looks of this place, the remoteness, the filth and squalor of my room, the administrative muddle, all this left me depressed. However, things can only improve. I think in a couple of days most of the administrative problems will get sorted out. Living conditions will get better. I will survive". (July 15)
>
> "For the last 3 days I have been pushing to get some necessary things done to my cell, and obtain some things I need, but such is the inertia and incompetence that nothing gets done. It is like flailing in thin air. I find that this causes my mood to fluctuate up and down several times during the day. Sometimes I get depressed at my helplessness, then some little thing gets done and I feel cheerful for a while". (July 17)
>
> "The big things one can handle, it is these petty little pinpricks that upset. All so unnecessary. All designed to just bother. They infuriate, and then when there is nothing one can do it leaves one with an awful feeling of helplessness. I haven't felt like this since the early days at Attock". (July 18)

"I suppose I will have to sweat this out too. But I shall survive. It is such a waste when a person's whole energy is used just to survive. The trouble is, if one spends too much in surviving, after some time it doesn't really matter whether one does survive because there's nothing left in oneself. But I think I can manage better than that. And perhaps it will not be too long. Anyway, knowing the worst, one can face it better. Now that I know where I stand with these chaps I feel better. I can outlast them". (July 19)

Bad as these things were, they were not the worst that I found myself wrestling with. What deeply bothered me all the time was my inability to understand why all this was happening to me; I could not make sense of it. Like Muslims generally, I had grown up with the unquestioned belief that the world was run by a just and benevolent God. My parents were not religious; they were content to live by the values and principles of Islam, and follow its ethical code; I adopted this attitude as well. As I grew up I did recognize that my rather simple faith did not deal adequately with many metaphysical and philosophical issues, but I was too busy living to worry much about such abstract matters.

From a young age I had believed that one had an obligation to live one's life in pursuit of some worthwhile goal. I had also believed my actions in December 1971 and afterwards were such as would meet God's approval. I took the resulting adversity that I and my loved ones suffered as a temporary phase, and had believed all along that a just God would ultimately provide the means to rescue us. In the last days at Campbellpur I had been thrown into turmoil when some verses I read in the Qur'an seemed to negate the basis of this hope. This was aggravated by my experiences in this new prison.

The various avenues of escape from my predicament were being closed off one by one. At the same time my conditions of imprisonment were becoming worse. The grounds for hope were being cut out from under me. I could not make sense of it at all. Having sacrificed myself (and my family) in the service of the just and merciful God who ran the world; why was he abandoning us? This tormented questioning went on all the time deep inside my mind; I could not find an answer. One day the feeling of being abandoned hit me particularly hard. I got very depressed; I knew that without hope for the future I could not go on. As I wrote in my journal:

> "I felt terrible today. I don't know why, but it was really bad. I got quite scared. At this rate I won't last very long. It was almost physical, though I know it really wasn't. I hope this was just an odd aberration. This solitary confinement and the bleakness of my prospects are getting me down". (July 25)

I continued wrestling with this existential conundrum. Finally, I worked out an answer of sorts, a possible solution. Something I could try out and see if it worked.

> "This hypothesis makes some sense. I now wait to see if future developments conform to it or not. If not, then I shall be thrown back into what I can only call chaos. But for now I find my turmoil at an end, I am at peace again. There is a sadness, of course, and loneliness. The solitariness of my confinement is a heavy burden. But all these are supportable, are bearable, if one is spared the agony of a world without meaning, a fate without purpose, a life without significance, and a future without hope". (July 26)

"Otherwise, life goes on. It has settled into a routine, and I go round and round like the ox tethered to the water-wheel.......... I remembered something I had learnt in the early days at Attock: *Life in prison is a succession of small victories and big defeats.* It is these small victories that enable one to live on.......... That one bad day has not been repeated. I have recovered my previous stoical condition. The worry at the lack of news from Jamila, and what may have caused it, only leave me braced to meet another blow. There is nothing I can do except take what comes. And try to survive". (Aug 1)

At long last, on the 3rd, Jamila arrived for a visit. Dani had driven her down from Lahore, along with Afroze and Yawar, and they were going to stay at a rest house for 3 days. I was seeing Yawar after quite a long time; he was looking older and had grown taller. It was such a joy to see them all and spend some time with them. The Superintendent, as usual, tried to be difficult; he said he would allow a second meeting only after a week! It took a lot of arguing to get him to agree to 3 visits over the weekend. Jamila told me that she had met some of the other wives, and many of them said their husbands were in quite a bad way owing to the strict conditions of their confinement (similar to mine, and including complete segregation where there were two in the same jail). However, before she returned she learnt on the phone from Lahore that orders were being issued to move us all to larger jails; I would be going to Multan. It was a great relief. Even though it would be farther away from Lahore, nevertheless conditions there couldn't be as bad as they were here.

The orders for my transfer to Multan Central Jail arrived on the 8th. They finally managed to arrange for a police escort to take me on the 14th. So, on the 27th anniversary of the founding of my country, a struggle to which I had devoted my youth, and in whose service I had spent my life, an armed police posse escorted me from the steel gate of one prison to the equally forbidding gate of another. Another grim journey, another prison, my fourth. So far.

11

Multan – The Outer Life

(August 1974 to March 1978)

[*I spent a little over 3½ years in Multan Central Jail. After about a month there my life settled into a routine with every day the same as the preceding one. Nothing much was happening in my life, either inside the prison or outside. In September 1974 I stopped keeping a daily journal and instead just noted down occasional reflections.*

I think it would be appropriate to cover my stay in Multan jail in two parts. This chapter deals with the conditions in the jail and the few significant events that occurred during this period – including the quite unexpected game-changing one towards the end of my stay. The next chapter reproduces selections from my journal, which deal mainly with my thoughts and emotions during this period. I gave this part of my journal the title "Wise Sayings of an Old Fool" (for reasons made clear in the first few entries).]

I reached Multan Central Jail, situated a few miles outside the city of Multan, at about 10.30 p.m. We (the escort and I) had to wait for quite a while outside the big steel gates while the jail officials were summoned to take over my charge. Except for some necessary exchanges I had kept quiet throughout the journey of several hours; the escort chatted off and on among themselves, and I noticed that their commander had quite a sense of humour. Now, as we waited outside the gate in silence, he turned to me and said: *To aap qaid barri sanjeedgi se guzaar rahen hain* ("So, it seems you are undergoing your imprisonment with great seriousness!"). I couldn't help bursting out laughing.

I was placed in a double-storied ward which had 4 cells on each floor. There was a high wall surrounding the yard, which was about 30 by 50 yards in size. The cells were in the usual poor condition and it took a few days for them to be patched up and whitewashed. I made one cell on the ground floor my bedroom and an adjoining one my bathroom, while the end one became my 'kitchen' where the prisoner-cook provided to me also lived. Gradually some basic furniture was provided, the kitchen got going, and, even though my cell was small, it was much more comfortable than my last one. The biggish yard seemed open and airy (there were a few trees in it), and, after some days, I obtained permission to take an evening walk in the main jail compound – there was an oval (earthen) road running around it, and this hour's outing was a pleasant change while the brisk walk gave me the exercise I was used to taking daily.

I found everything much more relaxed in this prison, mainly due to the Superintendent, who was a senior officer and seemed mature and balanced. The attitude of the warders was respectful and some even saluted me; little things like that make quite a difference in prison. My cell door was never locked, though the ward gate was kept locked with a warder on duty at all times; at night he was joined by a prisoner *numberdar* (trusty). I had got so used to sleeping with my door shut that the first night I felt like closing it, but decided against it (so as not to give them any ideas!). A few times some junior officer got itchy and started to tighten things up (once after a letter came from their headquarters requiring the rules to be strictly applied to us), but a word to the Superintendent would quickly put an end to this.

It was still solitary confinement, though: no one else was allowed to enter my ward (except on business). However, the ward next door was also double-storied (like most buildings there), and I could see the top floor over my wall; this helped break down the feeling of isolation since, even though I couldn't see anyone, I could sometimes hear people talking there or the radio playing. I was told a Nepali major was kept there; he'd been in this jail many years, probably for espionage. I learnt that he did a lot of gardening, and had made the ward green with flowers and vegetables. (Mine was dry and comparatively bare; no grass, a few bushes, no flowers). It seems gardening is a favourite hobby of prisoners; Nawab Akbar Khan Bugti was once confined next door, and also did a lot of it. The dacoit (gangster) Muhammad Khan was once in my ward, and also had a nice garden; that must have been quite some time ago because there was nothing left of it. The last occupant of my ward was, I believe, Mukhtar Rana. He probably did no gardening, but was said to have read a lot, mostly Mao Tse-tung!

A few days after I got there Wajid (Brig Wajid Ali Shah, also an Attock 'alumni') arrived. He was kept separately, but I ran into him in the Superintendent's office the day after his arrival, and we exchanged a few words. Even though we couldn't meet, just his being in the same prison cheered me up; I didn't feel as alone as before. Unfortunately, he didn't stay for long, and after a month or so moved to another jail. I learnt from Jamila on her first visit (shortly after I got to Multan) that the others were still generally in a bad way, mostly due to the rigorous application of solitary confinement and other prison restrictions in their jails. I was lucky in being able to cope well with the solitary life I was living, and also in no longer being subjected to the petty harassments and restrictions of prison life.

After about a month my life settled into a daily routine. After I got ready in the morning and had breakfast, I would sit down to my study of the Qur'an. In the winter this would be in the (barred) verandah fronting my cell, since it got some sunlight. I would work on this till lunch time. After lunch I would do some reading connected to this study (i.e., books on Islam) until evening tea. Then a most enjoyable walk in the central area of the prison; I did a brisk two miles in half an hour. Apart from the exercise this outing broke the solitariness of my day, since I saw other people around (though I never talked to anyone except to return some greetings). Then back to my small room and some light reading till suppertime, and often afterwards in bed till I dropped off to sleep.

As I have written earlier, in Campbellpur and Bahawalnagar I was buoyed by the hope that our case in the superior courts would ultimately free us; however, this was not going too well. After raising an objection to jurisdiction on behalf of Aleem Afridi and myself before the court-martial, Manzur Qadir, our counsel, had filed a writ petition in the Lahore High Court but did not pursue it. After the announcement of the court-martial's verdict and sentences he amended the petition and asked for it to be taken up. However, this request was denied and the High Court dismissed the petition without even a hearing. Manzur Qadir wanted to wait till later in the year (1974) to file an appeal in the Supreme Court, but before he could do this he fell seriously ill and had to be rushed to London. Unfortunately, this was to no avail, and, tragically, he died there soon after. That left our hopes seriously shaken as much of our confidence had been based on Manzur Qadir's ability.

The appeal was finally filed in the Supreme Court in March 1975 by M Anwar, a senior lawyer with a considerable reputation but past his prime; the reports of his performance when the case was heard were not very promising. Finally, the long legal saga came to an end when the court announced its judgement on Sept 14, 1975, rejecting the main contention in our petition (that Aleem and I were not subject to the court-martial's jurisdiction, and that consequently our trial had been illegal). However, it did hold that the court-martial did not have jurisdiction to try us on the second charge (of conspiracy to overthrow the government) and recommended that the government consider whether it would not be appropriate to reduce the single combined sentence imposed on us for both charges. Even though my hopes had dimmed considerably, this judgement still came as a blow (needless to say, I did not hold my breath regarding any action on the court's recommendation).

In December 1975 I submitted a petition to the Supreme Court asking for a review of their judgement in our case. It was based on the fact that some of the grounds which had been included in our case had not been dealt with at all in their judgement, and appeared not to have been considered. This was largely a *pro forma* gesture; I did not harbour much hope that it would come to anything. About a month later I got a letter from the court registrar rejecting the petition on the basis that it had not been signed and submitted by the senior counsel who had appeared in the case.

I then addressed an application to the Chief Justice requesting that these procedural requirements be waived on certain grounds that I advanced. I was quite surprised when in March 1976 the Chief Justice accepted my application and admitted the review petition, though he rejected my request to appear personally; instead, he offered to have the court appoint a counsel if I could not afford one. (Yaqoob Ali Khan had taken over from Hamood-ur-Rahman; he was a great friend of my uncle, Dr Shujaat Ali, and perhaps this was his way of making amends for his overly enthusiastic support earlier for rejecting our case!). Nothing happened after that for quite a while, which did not surprise me as Bhutto was still in power. After about 8 months the court registrar asked me to deposit some money so that the case 'book' could be printed. I did that and waited, but nothing more transpired.

With our legal challenge out of the way, soon after the Supreme Court judgement (in September 1975) Bhutto struck back. Under the prison rules a sentence of life imprisonment meant spending 14 years in prison. The government now issued an order that, in the case of court-martial sentences, "life" would actually mean imprisonment for life! Through another ordinance military prisoners were made ineligible for any remissions to their sentences. (Under the rules prisoners earned one month's remission of their sentence every year for good behaviour; in addition, remissions were awarded on special occasions (such as Eid or visits to prisons by dignitaries, etc) and these could add up to a significant reduction in sentences). Though this would not now affect me, it did have serious consequences for those of us who had received shorter sentences.

I now had to squarely face the prospect of passing the rest of my life in prison. This possibility had always been there at the back of my mind, but I had not paid it much attention while the hope lasted that Manzur Qadir would win the case for us. That was now gone. I buckled down to enduring whatever fate had in store for me. I knew I was strong enough to manage this solitary life for a long time, but, deep down, there was still a nagging worry about how long I would be able to last. I decided to take it one day at a time, and not think too much about the future. Two things helped me immensely in enduring my solitary imprisonment.

One of them was my ongoing systematic study of the Qur'an, to which I devoted several hours of every day beginning around the end of 1973 (and finally finishing towards the end of 1976). The study started off as a useful way to fill the considerable spare time my imprisonment forced upon me. I had always wanted to study the Qur'an for myself, but fully expected that my study would confirm the views of Ghulam Ahmad Parwez, a contemporary scholar of Islam, who had spent his whole life studying the Qur'an and whose views I found eminently sensible and acceptable. I had been introduced to his teachings in 1952 by my friend Ihsan (then Major, later Maj Gen, Ihsan-ul-Haq Malik; he was an exceptional person, and we remained lifelong friends).

As I have narrated earlier, towards the end of my stay in Campbellpur I had come across certain verses of the Qur'an that had very troubling implications for me personally. In Bahawlnagar I finally managed to come up with some kind of a solution to this dilemma that reconciled what was happening to me with my original belief; as my situation improved in Multan, this solution seemed to be valid.

Now, about a third of the way through the study, I again ran into a serious difficulty: certain fundamentals of the Qur'an's teaching (as generally accepted by Muslims, and which I had also unquestioningly accepted) did not appear to conform to human experience in the real world. This was the old conundrum of reconciling an all-powerful, just and beneficent God with all the injustice, suffering and misery that existed in the world. Some religions avoid the issue by putting off God's reckoning into the afterlife but Muslims do not take that position; for them God functions as much in this world as in the next. This problem was not just an academic issue for me, it affected my understanding of what had happened to me and what I could expect to happen in the future. Adding to the tumult in my mind was another dire implication: since this Muslim belief was apparently based on the Qur'an, this inconsistency raised the issue of whether the Qur'an was in fact divinely inspired, as it claimed and as Muslims (including myself) believed.

I found myself in a crisis. Not only was my whole belief system in question, so was the underlying significance of my actions and the resulting hope that at some point I would be helped out of my current predicament. I wrestled with this deeply troubling situation for two to three weeks in considerable mental anguish. Finally, almost indirectly, I had a sudden insight that solved the problem and ultimately led me to a new way of understanding the Qur'an's message. (I was looking up the term *khalifah* in Parwez's Qur'anic dictionary (it is a term applied to human beings in some significant Qur'anic passages). In discussing verse 2:30 Parwez argues against assuming that the word is linked to Allah (i.e., that the verse refers to Allah's *khalifah*) because that construction could only be valid in case of Allah's *absence* from the world. The word "absence" struck like a huge gong inside my head; it opened the door to a whole new line of thought and investigation. When I pursued this enquiry I found that the Qur'an repeatedly makes clear its position on this issue (namely, that God does not act in our world), which is very different from what I, and Muslims generally, believed).

From a merely interesting pastime my study became an exciting voyage of discovery into uncharted waters, and acquired a whole new purpose: I was breaking fresh ground, and opening up a new way of understanding Islam. At its end I had a fairly voluminous set of notes on the Qur'an, in which I dealt with all the main topics it covers on the basis of the Qur'an's own words. I hope to be able to publish the results of this study one day. Meanwhile, in Appendix 4 (page 216) I have included a piece written a few years ago in which I summarize the main features of these results.

The second critical factor that enabled me to survive the long years in solitary confinement was the support of Jamila and the children. Jamila kept in constant touch with me through letters and visits. She would make the trip to Multan once every month, flying in and staying for three days at a relative's place, and coming to the prison each day to see me. The jail was several miles outside the city and, since these relations did not have a car, she had to make the journey with whatever means were available: usually in a motor rickshaw, once or twice even riding pillion on a nephew's scooter. She would also bring food that she had had prepared in Lahore (for the lunches we had together, and some more for me to keep) and books and other things that I had asked for. These visits kept me going. Afroze also wrote frequently – long, chatty letters.

It was an immense relief to me that my family were coping so well with their drastically altered circumstances. Jamila and Afroze were living with my mother in Lahore, while Yawar initially remained in Karachi with Jamila's brother so that his schooling would not be disrupted. Jamila had never had to fend for herself, but now she was managing superbly. In no small measure this was due to the 'kindness of strangers'; as soon as people knew who she was they would tend to be extra helpful. Even Hanif Ramay, now Bhutto's chief minister in the Punjab, sanctioned a scarce and much-prized permit for her to buy a new car at the controlled price. This was in contrast to some of our more distant relations, who tended to play it safe. As did many of my army friends; with of course some notable exceptions.

On a visit from Karachi (where he now lived) our friend Gen Ihsan brought a gift of money from some of his rich friends there (whom we had also got to know); he insisted that Jamila keep it despite her protestations that she didn't need it as her brother Ijaz was sending her enough every month. Similarly, Ali Muhammad Dawood (one of the Dawood Seths, whom we had got to know in Karachi through Gen Ihsan) once came with the latter to visit me in Multan Jail, and insisted I accept a gift of money in spite of my oft-repeated reluctance. When Jamila visited Karachi a couple of times our friends there showed her many kindnesses, especially Parveen and Ahmed Khan. I had known Maj Gen Shaukat Riza and worked with him (one always worked "with him", never "under him") but we could hardly be called friends. One day he came to see Jamila and told her that he had recently retired, and was coming from the bank where he had just collected his commuted pension. Pointing to his briefcase he said it was all available if she needed it!

Much as I depended on my family for support, I realized that the prospect of my remaining indefinitely in prison required us to re-evaluate their future, especially the children's. There was nothing left for them in this country. Yawar in any case wanted to go to the US to study; the question had been whether we could afford it. In 1976 this dilemma was solved for us. Two of Jamila's brothers had emigrated to Canada and were living in Toronto; neither of them had any children of their own. They now offered to have the children come to Toronto and live with them, one with each, while pursuing their studies there. The children had already lived in Canada when we spent two years there while I attended the Canadian Army Staff College, and, even though they had been quite young (they were 6 and 2 when we went in 1961), they had happy memories of their time there. After the longish process of getting their visas, they finally left in the middle of 1977. Their last visit to say goodbye was a bitter-sweet occasion; while I was happy they were going, who knew when we would see each other again?

With the children going to Canada, and likely to stay there permanently, I began to seriously urge Jamila to do the same. There was no future for her here; I wanted her to also go to Canada and make a home and a new life there for herself and the children; it would be many years before they could fend for themselves. But she would not hear of it. We then decided that she should try and get me shifted to Lahore, so that she did not have to travel every month to Multan to see me; she could also visit every week if I was in Lahore. Under prison regulations prisoners were kept in jails near their home towns, and all the others in our group had finally achieved that; I was the only exception. She explored all the avenues available to us, and to family and friends, but each time was told that decisions regarding me could only be made by Bhutto. She even wrote to Begum Nusrat Bhutto, but to no avail.

Finally, a ray of hope appeared. Brig Muzaffar Malik had moved from the army into the civil service and, while working in the central government, had become a favourite of Bhutto; he had now been sent by him to the Punjab as chief secretary (presumably to keep an eye on things). Muzaffar's parents had known my parents, and the few times our paths had crossed in the service he had been quite cordial. When Jamila went to see him he was very sympathetic and promised to do what he could; he said he would run it by Bhutto on an appropriate occasion, and she should see him again after a week. When she did that he said Bhutto had raised no objection and he had issued instructions for the file to be moved for the transfer; it should be done in a week. When nothing happened in that time she went to see him again, and found him embarrassed and apologetic; apparently, late one night Bhutto had rung him up and chewed him out over the matter. It seemed that during one of his usual late-night drinking sessions Bhutto had recalled whatever grievance he had against me and vented his anger on Muzaffar (I suspect when I barged into the Governor's House in Lahore a few days after he had taken over, it probably threw a scare into him; he wouldn't ever forgive something like that).

Having exhausted our recourse to the highest court in the land there appeared to be no hope left of escaping the fate that had closed in on me. The only chance I had now was if Bhutto were to lose power, but there were no signs of that happening. He had, as I had originally feared, tightened his hold on power and established a virtual dictatorship. His clever and convincing advocacy, in words at least, of the cause of the common people had given him a solid base of popular support. He changed the rules to end the independence of the civil service, and brought it firmly under his control. To ensure his hold on the army, in 1976 he appointed Zia-ul-Haq as the new army chief even though he was the junior-most of the six candidates. Zia had first established his bona fides with Bhutto during our court-martial and he later assiduously built on this foundation to convince him that he was the most (personally) loyal general in the army.

Bhutto had nationalized many of the largest industries, including the banks, and, together with the government machinery, this gave him a huge source of patronage to cement the loyalty of his party members. He also dealt ruthlessly with many of his old colleagues who showed the slightest signs of independence. He was similarly brutal in his dealings with the political opposition, disrupting their activities, arresting and imprisoning their leaders, even having some of them killed. He had, soon after coming into power, set up a paramilitary force under his own control which he used as his special instrument of intimidation and suppression. Thus, even though more and more people were becoming disillusioned with his rule, especially among the urban population, lawyers and businessmen, he bestrode the country as its undisputed and unassailable master.

Under the constitution Bhutto had to hold an election in the second half of 1977. However, in order to undercut the opposition, he unexpectedly announced on January 7 that the elections would be held in March; nominations of candidates would close on January 19. Though he could probably have won a fair election, Bhutto wanted a two-thirds majority in the National Assembly so that he could amend the constitution. His administration and party activists went to work and there was open rigging of the vote. When the results were announced Bhutto's party had won 155 out of the 200 seats in the National Assembly to the opposition alliance's 36; most blatant of all was that the latter did not win a single seat in the metropolises of Karachi, Lahore and Rawalpindi, which were bastions of their support and where they had held massive election rallies. It is said even Bhutto was surprised and upset at the lopsided result, and berated his over-enthusiastic minions.

Naturally, the opposition refused to accept the results of the election and launched a public protest movement demanding a re-election. Bhutto used the police and his party henchmen in an attempt to crush the agitation, but it began to grow, especially in the big towns. Finally, unable to suppress it, Bhutto called out the army in Lahore. Even though the army resorted to firing on demonstrators for several days, killing many, the protests continued. Three brigade commanders whose troops were involved were very unhappy with this and repeatedly urged their division commander to convey to GHQ the general feeling in the army that this affair should be settled politically and not by using military force. When he failed to do so, all three of them resigned. Instead of accepting their resignations Zia immediately demoted and retired them on disciplinary grounds. But he could not alter the deep unhappiness within the army at what was going on, and its cause. Soon, Zia's corps commanders told him that Bhutto had to go; otherwise it would lead to serious problems within the army. He resisted for a while, but finally agreed. On the night of July 4, 1977 the army removed Bhutto from power and Zia declared martial law in the country. He assumed power and declared that another election would be held in October.

Within the short space of a few months the unthinkable had happened. It was something in the nature of a miracle that, though not entirely impossible, I could hardly have dared hope for. My initial reaction was mainly one of relief. No wild joy, no exultation, just a sudden relaxing of a long-borne tension, a feeling of lightness. This euphoria, such as it was, lasted only a couple of days. Then the public announcements of an October election suddenly raised the grim spectre of the possibility that Bhutto might again come into power. I was deeply worried by this.

Obviously, the same thought must have occurred to Zia, along with the dire consequences this would hold for him. He soon postponed indefinitely the promised new elections, and began to consolidate his rule. My fate was in his hands now, and we went to work to try and make him do the right thing. Jamila became very active, mobilizing the other wives to mount a campaign for the release of the Attock prisoners. The media was approached, as were persons of influence and those whose word might carry weight with Zia. Jamila and Shamim Afridi issued press statements and held press conferences, and soon the media took up the cause; editorials appeared, as did press statements from notables and letters to the editor. The leaders of the opposition alliance raised the matter in one of their meetings with Zia; he agreed that we ought to be released and said that this would have been done "if he himself had not been involved". Presumably, he meant that such an action would be taken as an admission that he had imprisoned us to please Bhutto. However, he did promise to rescind Bhutto's ordinance withdrawing remissions from military prisoners.

In early September Bhutto was arrested and charged with ordering the murder of a political opponent's father (the son, Ahmed Raza Kasuri, escaped the bullets fired at their car but the father died). His trial was held in the Lahore High Court, and by a strange quirk of fate the special prosecutors appointed by the government were lawyers who had represented us earlier – M Anwar and Ejaz Hussain Batalvi. Through them Jamila and Shamim attended one of the court sessions in October, and she wrote to me about it:

> "We got good seats quite near the accused's corner, surrounded by railings, upon which all eyes were fixed. At about 8.30 Bhutto came in, surrounded by police and military intelligence people in civilian clothes. The state he was in was worth seeing. He sat there looking like a caged beast, his eyes darting from left to right around the courtroom. Wherever he looked, people were staring at him, and he would then tilt his head back and stare at the ceiling, sometimes trying to smile or frown. We could see that he was taking a lot of interest in us as we were the only women among the spectators; I suppose he wondered who we were. He was looking thinner and his puffy eyes showed that he

had not slept much, but otherwise he had a healthy look. We were told that he wasn't getting any liquor in the jail; a couple of times his wife tried to smuggle some in but was prevented.

Ahmed Raza Kasuri was on the witness stand and when he finished giving his statement he was cross-examined by one of Bhutto's lawyers, who did a rather poor job. Strangely, Bhutto has no lawyer of any stature to defend him in this trial where he is fighting for his life (we had the top ones!). Once when Kasuri answered a question Bhutto quite audibly said, "Nonsense". Kasuri heard him and complained to the court upon which Bhutto stood up in a temper and said loudly, "He is talking nonsense". Kasuri got enraged and shouted at Bhutto, "Shut up and sit down!" The presiding judge then intervened and told Bhutto to sit down, which he then did quietly.

We were told that early on in the trial when one day his corner was newly fenced off with railings, he was furious and complained to the court that he was being humiliated and ill-treated. Chief Justice Mushtaq replied that he was no longer the Prime Minister but only an accused in a murder trial, and normally would be sitting on a wooden bench with handcuffs on; he should be happy they had given him a cushioned chair.

When we went to the court this morning we thought we would feel quite pleased to see Bhutto in the dock. Can you believe it, we actually felt sorry at his pathetic state, his misery and helplessness. What a fall!"

Finally, in January 1978 Zia repealed Bhutto's ordinance which had denied military prisoners in civil jails the normal remissions they were entitled to under the rules. This had an immediate impact on most of the Attock prisoners since, with the accrued remissions, their sentences had already been completed. Most of them were soon released. That just left Aleem Afridi and Farouk Adam (in Haripur Jail), Asif Shafi (in Kot Lakhpat Jail), Nadir Parvez (in Lyallpur Jail) and me still in prison. Since Zia seemed to show no inclination to do anything about us, the only avenue left now was to pursue the legal option. In November 1977 I submitted a request to the Chief Justice to fix an early hearing of my petition, but got neither a response nor any action.

We had also been making efforts for my transfer from Multan to Lahore, and finally, in March 1978, I was moved to Kot Lakhpat Jail, a few miles outside Lahore.

12

Multan – The Inner Life

Journal Excerpts : Wise Sayings of an Old Fool

On reaching the last quarter of one's expected life-span, with the children coming ever closer to the threshold of independent life, there is a strong urge to transmit to them whatever little wisdom one has accumulated. So that their way becomes easier, their path a little smoother.

However, such a transmission of wisdom is neither easy nor smooth. For, unfortunately, to one's children, when they reach this age, one is an old fool with nothing worthwhile to teach. It may not be formulated in those words, or even thought in that form, but this is the sum-total of their view of the matter.

So the old fool is for the most part left to confide his wisdom to his notebook. Along with sundry thoughts, views, recollections and reveries.

* * * * *

I must confess there is some poetic justice in this. I do not recollect my father trying to pass on to me any accumulated wisdom, but if he had I probably wouldn't have paid much heed. In my teens I began to feel fairly self-sufficient in the matter of wisdom. Or looked for it in books and among the few contemporaries I considered intelligent.

I suppose it has something to do with the generation gap. Young people feel their parents can't possibly be in a position to give worthwhile advice about their contemporary, modern, brand-new problems! Forgetting that human problems don't really change that much, and that their parents have been there before them.

This is one advantage the old have over the young: that they have been young once, while the young have not yet been old. But the young seldom concede this. Having been young once is very different to being young now, they think. It is not, of course, if one uses a little memory and imagination.

* * * * *

But if one is to transmit wisdom, how is it best done? One day, walking in the yard of my prison ward it occurred to me that one should be able to distil out of one's experience a certain number of pithy aphorisms. This was many months ago and, if I remember aright now, at that time all I was able to dig up was one. Though certainly pithy I am not sure how wise it is. Anyway, I reproduce it below. Doubtless I shall think up others off and on, and will record them for posterity!

* * * * *

In human affairs there is no such thing as Never. Nor is there any such thing as Always.

* * * * *

Still, there are many things which cannot be captured in an aphorism, pithy or not.

Some time ago in one of the letters from home mention was made of my son (16'ish) having bought some Harold Robbins book. Aha! I said to myself, it's time we had a chat on the subject of literature versus trash.

So, a chat not being possible under the circumstances, I wrote him a longish man-to-man letter. Not at all didactic or patronizing or anything of the sort; I consciously made an effort to avoid sounding as if I was trying to impose my standards upon him. I simply said that time was always short and, though occasionally one read books mainly to pass the time, there were many others from which one could derive both pleasure and profit (speaking of fiction only). The good writers had this quality: not only did they amuse and divert and entertain, or grip and move and shake, they also had something serious to say. Something about life and human nature, and our place in it all.

The reply I got was, in effect: thank you very much for your views, but I have my own on the subject; I read for pleasure, not profit.

* * * * *

Months passed. The urge grew again. One day in a letter my daughter (21'ish) asked why there were no good books on Islam for children. (She was thinking of her children-to-be, I of mine-that-were).

The subject seemed ideal. It bracketed all the essentials, all the fundamentals. Neither I nor my wife has thrust any religious instruction on the children; mainly because we were never too sure ourselves. But in these last few years in prison I had been studying the Quran, and I thought I had discovered most of the answers it gave to the eternal questions. They seemed good answers, and I thought this an excellent opportunity to convey them to the children.

So in reply I wrote what was a preliminary letter, trying to explain what religion really was, or should be, about. And that perhaps I was now in a position to provide answers to many of the questions involved. I expected a reply all eager to learn these answers.

What I actually got was rather different. She didn't agree with what I thought religion should deal with. The "ten commandments" were all the religion she needed, and she didn't think there was much need or justification for God!

So that particular piece of wisdom also didn't get very far either.

* * * * *

The pity of it is I think older people really can help the young. We have already been where they are now, where they will be next year, and the year after....... Times change, true enough, the world changes too, but human beings change very little, nor do the problems they face. And in any case the changed world, the changed times, are not all that changed. We have seen these changes too, we have lived through their transmutation; it is not as if we're Rip Van Winkle reawakening after a sleep of centuries.

It is not so difficult, with a little imagination, a little sympathy, to picture the situation in which the young find themselves today. And to tell them a few things which could make their journey easier, lessen the number of bumps they hit on the road.

There are, of course, some things that cannot be taught. How can you possibly make a young person in love believe that this intoxicating ecstasy does not last, cannot last?

But there are many others that could be usefully passed on. If only someone were ready to take them aboard.

* * * * *

Life was good when I was in college. One had just stepped across the threshold into a huge, wonderful world. Home was nice, but it was also nice to be on one's own in a big city like Lahore. Living in the hostel was fun, even though it wasn't at all plush. Pleasures were few and simple, but one relished them. Companionship. Movies. Listening to film songs. Smoking the odd cigarette. And above all, books. The library of the college was an Aladdin's cave for me.

I still recall the pure ecstasy of sitting in my room in the Quadrangle on a cold winter evening, wrapped in something, the table lamp casting a pool of light on the Dumas novel I was devouring (I had discovered a complete set of his works in the library) and, to top it all, I had a plate of *gajr ka halwa* (carrot 'fudge') from the canteen. I ate it morsel by morsel as I wandered the dark streets of medieval Paris and the corridors of the royal palace, mingled with Kings and Queens, princes and courtiers, knights and captains, clowns and jesters.

The whole world lay spread out in front of me, at my feet. There was nothing I could not achieve. And I was in love. It was all so wonderful!

And all on 20 rupees a month, which was the pocket money I used to get. The canteen bill (mostly breakfast) came to about 12 to 15 rupees. And one lived on the remaining 5 to 8 rupees. I don't think I have ever been so happy and carefree again as I was in those days.

While at college I saw lots and lots of movies; even though one didn't have much money to spend, they were so cheap. To see a matinee all one needed was 11 annas – 2 annas for tonga fare and 9 annas for the movie ticket (a rupee was 16 annas). Four of us would get together and, in the sizzling heat, hire a tonga and go to the movie house (Regal or Plaza mostly). I don't remember if they were air-conditioned then, but they were certainly cool. By the time the film finished it was evening, and we could stroll back to the hostel along the Mall.

Another great institution was the morning show. These were also cheap, and usually re-runs of good old films. The only problem was the terrible crush at the ticket windows; no one believed in queues in those days. Often the person chosen to get the tickets would put on some oldish clothes to cater for the likelihood that they'd get torn. I have seen people launch themselves on the heads of the crowd around the booking window, and 'swim' on top of this mass of people up to the window. It all seemed great fun.

It was through morning shows that I discovered Greta Garbo. She was wonderful. Many of her pictures came as morning shows and I religiously saw them all. The same sort of impact was created later on by Ingrid Bergman. These actresses were in a class by themselves.

We also used to see quite a few Indian films. Their standard had gone up considerably around 1943-44; some excellent pictures were being made by the director Mehboob. Then there were a lot of light ones meant to entertain, which they really did. An almost standard fixture in all of them was the dancer, Cuckoo, who showed a commendable amount of leg in her dances. In those days many pictures I saw featured an actress named Shobhana Samarth. At the time I thought she was undoubtedly the most beautiful woman in the world!

* * * * *

I suppose one can get used to prison, as one can get used to almost anything. I have got so used to sleeping with a locked door, that the unlocked one now makes me feel insecure. I nearly closed it the first night but then didn't so as not to give these chaps ideas.

When I get out I think I will find it difficult, initially, to get over the absence of security, solitude, and most of all, the lack of problems. One has piddly little problems in jail, and all one can do is to keep on reminding and bothering people to do something about them. Not problems which require one to think of alternative courses of action, choose one, and then follow through on it. It is going to be quite overwhelming at first to learn again the business of coping with life.

But these are trifles compared to the great gift of freedom one would get in the bargain. We take our freedom so much for granted. We fret and fume when some of that freedom is restricted by people or circumstances which prevent us from doing some particular thing, but forget how much (and much the more valuable part) of freedom we still have. We can for the most part go where we like, do what we like, speak to whom we please, meet whom we want, live our lives our own way. Despite all that hedges him in, the poorest of men is a thousand times more free than a king in prison.

Sometimes, sitting in my yard in Bahawalnagar, I used to envy the sweeper cleaning out the drains because I knew he'd soon finish his work and walk out of the gates – a free man.

Looking out through the bars of a prison one's eyes, searching for heaven, do not have to travel very far. The memory of one's home and family is heaven enough, indeed all the heaven one can take. But this construction is also all wrong. For beyond the bars are the walls, the inevitable, unavoidable, ubiquitous prison walls, and looking out through the bars they are all one sees; walls that so completely cut off the world that I could easily doubt it exists at all.

* * * * *

"Simplify! Simplify!" Thoreau once said. When one comes down to it, or much more usually, is made to come down to it, it is remarkable how much one can simplify.

For two years now my world has comprised a small room, about 12 feet by 7 $^{1}/_{2}$, with a barred verandah, about 7 feet by 7 $^{1}/_{2}$ feet. With the few necessary items of furniture in the room there is barely any space to walk about. Out of the 24 hours of the day I spend about 22 in it. In the morning I am in the adjoining bathroom for about an hour all told, and in the evening I go out for a brisk hour's walk in the main prison yard. During the winter I spend the mornings in the verandah, which catches the sun. For the rest, my little room is my entire world.

Small though it is I do not feel confined because the iron-barred door in the verandah is open, except when I choose to close it from inside. And because I can, in my mind, wander the wide world, walk back through the centuries, even venture into the mists of the future.

Except for my prisoner-cook I seldom talk to anyone. Even with him it is hardly more than half-a-dozen sentences a day. One difficulty is that he speaks a dialect I have problems following; he probably finds my language equally difficult to understand. While we manage all right as far as business is concerned, general conversation inevitably suffers.

Yet I do not feel lonely. Because of books. Reading a book is holding communication with a human mind, another human being. And because the books are selected, they are very fine, very interesting minds indeed. It is because of their company that I do not feel lonely.

My life is simplified in many other ways too. Meals comprise the same one dish for lunch and dinner, 5 days a week, and another dish on the remaining 2 meatless days. Week in, week out. Identical. I probably eat with a better appetite here than I ever did before.

Two sets of clothes – *kurta pyjama* – meet my clothing requirements. It is only when I have to go to the office, or have visitors, that I take out a different set. And the same applies to all the other impedimenta with which we usually clutter up our lives.

* * * * *

I said I do not feel lonely. I speak about the company of human beings. But there are of course persons who are much more than that, persons who are really part of one's being. To be separated from them is to have lost a part of oneself. It is more than loneliness; it is an amputation that can kill. If one survives, it is as a cripple.

In my case the missing parts are my wife and children. But I have been lucky. Sometime in those early days of April 1973 when my world suddenly collapsed upon me and the tearing wrench had occurred, I managed, somehow, to cauterize the raw bleeding wound, to switch off that part of my mind that felt the pain. It remains switched off; in that whole area there is now just a numbness.

I suppose maintaining this numbness must be exacting a price. Whatever it is, it is well worth paying.

* * * * *

Sometimes one comes across a book which seems to have been written especially for one. Every word is a delight. I have just been reading such a one – Anthony Burgess's "Enderby's End". It is not a weighty or profound book, but it is sheer, unmitigated pleasure.

Others in this category are Hemingway's "For Whom the Bell Tolls", and James Gould Cozzens' "By Love Possessed" and "Morning, Noon and Night". Each word seems specifically written to strike deep chords of pleasure and meaning and response inside me. With them it's not so much reading a book as undergoing an experience. The experience of living inside the book, inside its protagonist, and, sometimes, inside many of its characters.

Hemingway is probably my favourite author. The things he writes about are things that I understand. Not the exterior events, but what is happening to his characters inside, and the hidden forces of circumstance that inexorably lead them to their destiny. The hallmark of the Hemingway hero is grace in the face of inevitable defeat, if not doom. This strikes a chord deep within me; it's something to which I respond strongly. Ever since my college days, when I first discovered Hemingway, I have found myself identifying with this hero.

From those early days this theme of a man struggling against fate has moved me greatly. I do not know why; I had never visualized myself cast into such a role, at least never consciously. And as for intuition, for a long time it was the other way around – I felt I would be able to achieve a great deal; I had a certainty of success. This gradually withered away, but even before that happened I found that I was powerfully affected by this theme. Gradually the Hemingway hero became my ideal: grace confronting inevitable failure, if not destruction.

What drew me towards Hemingway also was that he writes about reality, and he writes with truth. The reality he sees in the world is the reality that I see, and what he says about it sounds to me true. What he writes about is man's struggle in life, a struggle which seems to end most often in failure. But not defeat. His heroes fail, but they are not defeated. As he has his Old Man say: *A man can be destroyed but not defeated.* And sometimes he writes of the wreckage that is left behind.

His vision of life, and the way he writes about it, move me deeply.

* * * * *

Nothing is as good or as bad in actuality as it had seemed in prospect.

* * * * *

There are some phrases (often involving a paradox) that are too clever by half. They sound very nice – witty, pregnant, even profound – but on closer examination don't actually mean too much. One such phrase is: life in death. Another is its reverse: death in life.

Yet I find that these two phrases rather accurately describe my present existence. It really is life in death; and also death in life.

For most people who ever knew me, or even knew of my existence, I am as effectively dead as if buried under six feet of earth (instead of behind several times that of walls).

Even for my immediate family I am practically dead. But a rather nice kind of death, as deaths go. I can be written to, and replies come from the grave. For a few days every month or so I seem to come alive, and can be visited, and seen and talked to. (For my children it's usually at much longer intervals).

Not a bad death, really. Someone once said, "The trouble with the dead is that they're so awfully dead". I'm dead, but not too awfully.

* * * * *

By any criterion the life I am leading is a most peculiar one. Since I manage to lead it quite successfully I must be quite peculiar myself. Or, rather, have become so.

I do not find in myself too many visible signs of peculiarity, though. A tendency to talk to myself, but only a few sentences at a time. More often to exclaim or expostulate loudly. Sometimes address a question or a remark to a picture on my wall.

Then there is the mildly annoying physical problem of stomach acidity that appears to have a psychological cause. Other than that I haven't noticed much else. I sleep fine, no strange dreams or nightmares; I eat well, work normally. I do hope, underneath, it isn't a can of worms!

* * * * *

Among my contemporaries there weren't many I was close to. Most of them are now dead.

Within the family the only persons my brother Kaiser and I were close to were our cousins, Zulli and Iffi, also brothers. We were almost like four brothers, of similar age, growing up together. Iffi, who was about my age, died in 1947 in a car smash; he was about 18. In 1952 Kaiser went in another car accident; he was 22. Last year Zulli died; he was only a couple of years older than me.

In my college days I really made only two close friends. Qasim died about a year ago. For Riaz I died in 1973, so that amounts to the same thing. Sometimes one feels like a ghost wandering around. But I forget that I am dead, too.

* * * * *

When one is young death seems very far away. When someone died one was suitably awed and impressed. And if it happened to be someone close, one felt grief for a while.

But when one reaches the last stretch, death is a much more familiar figure. Its shadow is a constant and from this vantage point it's clear how everyone, young and old, is in fact walking in this shadow. Many of course are unaware of it, as once one was oneself.

Nowadays when I read or hear of the death of someone I know, my reaction is similar to that of a popular Urdu newspaper. Whenever someone newsworthy dies this paper prints a minor editorial under the heading (for example): Ah! Qasim Rizavi.

* * * * *

All human beings are fallible. Everyone deserves a second chance. And, if possible, a third, and a fourth, and a ……

* * * * *

When someone does something bad, we brand him a bad person. If someone does something foolish we consider him a fool. But how often do we ourselves do bad and foolish things, and yet know that we are neither fools nor bad?

* * * * *

How much of our view of ourselves is merely the reflection we see in the mirror of other people's attitude towards us?

Can one really retain a high opinion of oneself if the rest of the world persists in treating one as dirt?

The converse I know is true. To be nothing, to have nothing, and yet to be treated as something helps considerably in retaining some self-regard. Especially in prison.

But does this business go beyond the sphere of opinion? If you treat someone as an honest person, will it merely make him or her think of themselves as honest, or will it tend to make them more honest?

* * * * *

Does my sub-conscious reside in my stomach? The other day I was told something which upset me considerably. But afterwards, when I thought about the matter for quite some time, there was no longer any feeling, any emotional reaction. Consciously, I was not upset any more.

After a while I became aware of an ache in my stomach, and then of a severe headache. A headache of the type I sometimes get, quite rarely really, due to stomach acidity. The headache remained till the next morning.

When I realized what was happening, I recalled the time when this stomach problem had commenced. It was in 1972, when my time of troubles started. In those days when the court of inquiry was going on – a period of two months – I never once felt upset or dejected or worried. I was rather surprised at how well I was reacting. But it was in those days that this stomach acidity problem started. At the time I did not connect it to what I was undergoing. Later on, of course, I did.

But this is a new angle. It is a commonplace for stomach acidity to sometimes be caused by worry, stress, tension, etc. What seems to happen in my case is that worry, stress, tension, dejection, etc all slip through the mesh of my conscious mind, leaving little trace there, and lodge in my sub-conscious. And all that this load of submerged trouble seems to produce is acidity in my stomach!

Considering the nasty tricks an overburdened sub-conscious can play on a person, I am very lucky that mine seems to vent all its spite on my stomach alone. Antacids are a cheap psychological remedy.

* * * * *

We age so imperceptibly that we are hardly conscious of it. When we reach the stage of growing old the changes must be fairly obvious to others who see us occasionally, but to ourselves, who see ourselves every day, they are hardly noticeable. The new wrinkle or gray hair passes unnoticed. I suppose the same is the case with other aspects of ourselves.

I am now nearing 48, and therefore am growing old. In appearance the changes are not too noticeable – to me, at least. The wrinkles not too obvious, except around the eyes. A slight sagging of the face. My hair is still mostly dark, though there is some gray hiding behind the ears.

I find exercise no problem. My 2-mile walk every evening is done in 29-30 minutes without effort. But there is an increase in stiffness; bending down or squatting and then straightening evoke a muffled groan, sometimes an invocation of God.

For some time now I have also noticed a mental change, a sort of absent-mindedness, which wasn't there before. When I am doing some routine tasks and thinking of something else, I find I have done something wrong (for example, replacing a jar in the cupboard I suddenly notice I have placed it where another jar always goes, and not in its usual place). I had no awareness of doing this. This was never the case previously. But otherwise there seems no impairment of memory. I suppose it is all a part of this business of growing old.

* * * * *

For most people the thought of dying before one's time is an unpleasant one. There is so much in life that they would have to give up. Thinking of others, they can see how their own death would hurt and harm so many other people.

However, since the possibility of sudden death always exists, this becomes a source of some worry and anxiety. More for some, less for others. Perhaps only sub-consciously for many.

Then there are those unfortunate ones for whom life is so painful that death seems to offer a welcome reprieve. They wait for death with open arms, or go forth to seek it themselves.

I don't think there would be many for whom life and death, living and dying, offer equally attractive prospects. I find that my circumstances have placed me in this rare category.

Life is not so painful that I would welcome death, but I can face the prospect of dying without any sense of loss. The life I would give up is not worth too much. As for my loved ones, I do not think my dying would be much of a loss to them either. Of course, there would be temporary grief, but I am already more than half-dead to them. Perhaps there would, after a while, even be relief at not being tied to a half-corpse in a half-grave. Materially, too, my death would make no difference to them, since I can no longer support them. Released from this tie they could start living their own lives.

Sometimes the thought of death is a delicious prospect of rest after a struggle that has gone on too long, and lost its point.

* * * * *

What I wrote in the previous entry needs some qualification.

My life is not utterly useless in these circumstances. The work I am doing these days is, I think, of some value. It may have no practical outcome, but in itself, I think, it has significance. If nothing else it gives me the feeling that I am, even in prison, engaged in something useful, something creative.

My death would probably make a considerable difference to my wife. At present her life revolves around the children and me. Within a year or so, if things work out, the children will have gone to Canada. Then the object of her life here will just be me. If I go, there will be a great emptiness for her. Even while in prison I think I provide a centre around which a large part of her life is organized. This is the way things are in our society. There is little opportunity or incentive for women to live their own lives. Their lives are usually built around those of others.

I had suggested to her that she go to Canada with the children. But she will not countenance the idea. What will she do there? How will she support herself? she asks. The latter is a problem, otherwise she could make a home there for the children.

I am now trying to urge her to develop a life of her own. Take up with some old friends again. Start doing something to supplement her income.

While I realize that my dying would create a sudden problem for her, she could overcome it by beginning to lead her own life. Perhaps in the long run it would be better fcr her. It is not healthy to live a life centred around a grave.

* * * * *

Another reason people are afraid (or anyway reluctant) to die is because of uncertainty about what happens to them afterwards.

For me this, too, holds no terrors. If there is nothing beyond death, so be it. One has made a mistake in living as if there were. But that hardly matters, for a life that is only a momentary spark in the all-encompassing darkness is in any case meaningless, pointless. What does it matter how one lived it? What does it matter if it ends today instead of after another 10 or 15 years? Let it end, the sooner the better. Jokes in bad taste should not be unnecessarily prolonged.

If, however, there is a continuation after death, then I am ready for it. I think I can face an accounting of my life.

There is nothing morbid in these thoughts. I use up the life I have, such as it is now, in as constructive a fashion as I can devise. I shall continue to do so. I have no hopes for the future, of any sudden change for the better in my circumstances.

I do not desire death. I do not dwell on the idea of it. But sometimes a great weariness overcomes me, and the prospect of release from this prison of life, of surcease of this continuous effort to breathe in and breathe out, of laying down this burden now carried for far too long, appears terribly sweet. Peace and rest, at long, long last.

* * * * *

In Lebanon a Palestinian refugee camp has been under siege for over a month. There are women and children in it. They die in the fighting. Their underground shelters collapse in the bombardments and bury them in the rubble. For God knows how long the water supply has been cut off. Children are dying of lack of water. There is no medical aid; the wounded lie unattended till they die.

In the whole of the great Arab world, in the whole of the great Muslim world, not a voice is raised. Governments and peoples that cry themselves hoarse in support of the Palestinian cause now utter not a whisper. Why? Because politics is behind the butchery. Because the killers are Christian Arabs. Because supporting the killers are Syrian Muslim Arabs.

Will it be any surprise if these children – those that get out alive – when they grow up plant bombs, hijack planes, kill? What do you do in a world that is a cruel enemy except kill till it kills you?

* * * * *

There is an image stored in my mind of a man standing in front of a wreck, looking at his hands, and saying, "What have I done? What have I done?"

It is not some black crime or monstrous folly that he has committed, but some little carelessness, a few moments' thoughtlessness, a small inadvertence. But its consequences have been catastrophic, for beneath the wreckage lies his family.

Perhaps he was repairing a wall and, without thought, pulled out a brick or two and the whole house collapsed, burying his family beneath. Perhaps he was driving a car, and for a few seconds his attention wandered, and now he stands besides the smoking wreck of the car within which lie the broken bodies of his loved ones. And he holds out his hands, looks at them, and says, "What have I done?"

I do not know where I got this picture from. Was it some film? Or a book in which I read about some such scene? Or is it a creation of my own imagination?

It is an image that will not leave me. For the man in it is me.

* * * * *

The image I last wrote about is dramatic, but nevertheless true. Within the twinkling of an eye my family's whole world crashed about them. No, not quite – there were two twinkles, two crashes. The first was in August 1972. Without too much trouble, we rebuilt a world of sorts. It was mostly stage scenery, but it gave the illusion that we had rehabilitated our world. Then in March of 1973 it crashed irretrievably, totally, fatally.

For the children it is perhaps not too bad. Their lives lie in the future. Perhaps we will be able to salvage their futures from the wreckage. If so, they will not have lost much.

But my wife has lost everything. The time had just come when, materially, we would have reaped the fruits of the deprivations, the relinquishments, the sacrifices, the hard times of the past. For her the present and the future have become much harder than the past ever was. It was to be a time of ease and fulfillment. And achievement, too, for wide vistas would have opened up. It has all become dust and ashes. For her there is no bright new future waiting to be built. Only the dead weight of a wrecked past to be dragged around.

Sometimes when she has come to see me, and been unable to get transport, she has come to the jail riding pillion on a scooter. Riding thus through the streets of a town where she once had every right to expect she would one day be chauffeured around in a limousine. She has had to put up with the rough tongues and rougher malice of people who would have, under another dispensation, crawled on their knees before her. One could go on and on. But each succeeding cup is as bitter as the previous one.

Not once has she mentioned this aspect of things to me. Nor have the children.

But I still look at my hands, and still ask myself the same question. Not too often now. How many times can one go on asking a question to which there is no answer? Asking it does not hurt too much now, either. During that terrible first month at Attock I tasted to the full the bitterness of this particular poison, felt to the full the thrust of this dagger. However, the cauterized flesh soon loses all sensation.

* * * * *

The fool, emerging from his house in the morning and finding a lion cub playing in his courtyard, gives it a couple of kicks, saying, "No better than a stupid puppy!" This he does daily for a time.

The wise man, confronted with the same thing, will either get out his gun and shoot the cub. Or begin feeding it milk and meat every day. Or he will pack his bags and find another place to live.

* * * * *

When, after a lifetime spent serving as fuel in the furnace of life, all one has to show for it is a few drops of supposed wisdom, distilled in the retort of one's mind, there is inevitably a great desire to give them to someone to use, rather than tossing them away.

* * * * *

What matters in a human being's life is not what happens to him or her, but what he or she does. Or fails to do.

* * * * *

How important it is that human life on this earth is short. For, otherwise, how could the tortured endure his torture, the miserable his misery, the proud keep from bursting, the powerful from going completely berserk?

* * * * *

For most of us the pain and suffering that afflict us in our lives are no more than the pain and suffering that we have inflicted upon others.

* * * * *

The fatal flaw in the habitual liar is that he starts to believe his own lies.

* * * * *

Essentially, our fate is what we are. External circumstances merely provide the occasions for the unfolding of the destiny imbedded in our being.

* * * * *

Many of us, perhaps most of us, live on hope. But hope is only another name for not knowing the future. If we could know the future, how many would consent to live on?

* * * * *

If we do not know the future because there is no fixed future to know, that is bearable. But if we are denied knowledge of an unalterably pre-determined future, that would be a cruel joke.

* * * * *

The conversion tables give no factor for changing distance in miles into pain in the heart. How could they, since four thick feet of wall can cause as much sorrow of separation as two hundred long miles of space?

(Eid ul Azha, 1976)

* * * * *

Much of what has happened to my family would have happened had I died. In my adult life there have been several occasions when I have missed death, or death has missed me, by only a few inches or seconds. And there must have been many others when this has happened without my realizing it.

But that would have been an accident............ Is this any less an accident?

On the other hand, in each of those possible mishaps there was as much culpability in terms of my carelessness or foolishness as in this actual mishap.

There is no consolation to be found in this direction. Is there no way out of this recurring agony?

* * * * *

December 19 – an anniversary of sorts. The fifth anniversary. How different the same event appears from different points in time.

In prospect, it was something that had to be done. The anger inside me had to be expressed in some way. A white-hot anger. There was no violent emotional storm that drove me to it, only a cold, implacable fury. It was also expiation for my own guilt. For 20-odd years of compromises, of dalliance with noble purposes, of waiting upon circumstance. It was a way for the Army to wipe off some of the shame that bespattered its face. It was the least one could do for those who had given their lives in the past 3 weeks. But there was more to it than that. It was a way of averting a very real danger that loomed ahead. It also offered some hope for the future after the shame and squalor we had gone through.

Very clear in prospect, too, was the personal danger involved. In the middle of the road ahead stood Death as a very live possibility. But by then I had come to terms with death.

The event itself was strange. The first step taken, it was like having mounted a roller-coaster and pushed off. It plunged ahead with its own tremendous momentum. One just held on, and for form's sake made gestures as if one were driving. As in a roller-coaster there were terrifying plunges and exhilarating ascents. But above all, the unstoppable rush forward. Everything worked, everything went right; nothing could halt it; the few who tried were brushed aside like flies. It was as if a huge gear-train had already been assembled, and was lying there waiting for one button to be pressed. Later on, looking back, I was to feel that day I had truly been in the hand of God.

In retrospect, at first, just surprise. A pleased surprise that it had all gone so smoothly, so easily; that I had come out of it alive. A tired relief that I had managed to pay off some of the debt. Then, a burgeoning hope. For a while it seemed that the expectations would in some measure be realized. At that time this act seemed the high point of my life; the one solid achievement.

Later on, dust and ashes. Far from having averted a danger it now seems at least possible it made the longer-term outcome worse. It is probable that what has happened would have happened anyway; I only made it happen sooner, and removed some possibilities of alleviation. Yet at the same time I am fairly certain that what I did saved turmoil and bloodshed; how much one cannot say. There are at least some persons alive today who would otherwise have been dead.

For myself, what had once seemed the zenith of my life now appears to have been the commencement of the nadir. For my family, my loved ones, unmitigated disaster. All flowing directly from that December day. On my conscience, several other shattered lives.

A bitter anniversary. And yet, as time moves on the perspective keeps changing. Will it change again for that 19th of December?

* * * * *

Perhaps the secret for enduring suffering and affliction is perspective. A human life is but a blink in the unwinking radiance of eternity. What matters the momentary pain and sorrow that fills it if all else be peace and joy?

And even if there be no eternity, no joy, no peace, perspective can still reduce it all to meaninglessness. What significance has the span of one life when set against the billions of years we know the universe has existed, the billions that still lie ahead? What matters one human being among the billions that have, and do, and will exist?

But these are mere words. Of time, whether it be measureless eternity or computed eons, all I really know is the sixty or seventy years of my conscious life. Of the billions of beings the only one wholly real to me is myself.

In the final analysis, when all else fails, it is a case of gritting one's teeth and enduring. Where possible, fighting. Because one is human. Because there is a certain pride in being human.

But when one is stretched beyond the limit, when one begins to lose one's humanness, it is time to quit – as gracefully as one can.

* * * * *

In my study of the Quran I found it confirmed our observation and experience that God does not appear to function as a causal agent in our world.

This doctrinal confirmation of an empirical suspicion naturally evoked several reactions. There was the personal implication involved: it just about extinguished the last little flicker of hope.

But by far the strongest reaction was a feeling of relief at being able to absolve God of the responsibility for much of what happens in the world, of all the injustice and evil, the pain and suffering, the misery and squalor. It was as if a great oppressive burden had been lifted from my mind.

* * * * *

One of the saddest things one can experience is to be whole in oneself and watch one's body and brain slowly deteriorate.

* * * * *

Even when a thing is 99 percent certain the wheel of chance still contains a space labeled 100. This fact affects all our lives. In prosperity it precludes complacency. In adversity it can prevent despair. In between it adds a certain zest to living.

It allows even the hopeless to sometimes hope a little.

* * * * *

Often, our sins against our fathers are visited upon us by our children.

* * * * *

A young person with seventeen years of life and twelve of schooling cannot afford the fatal luxury of waving banners with strange devices and setting off on crusades under them.

* * * * *

The image will not go away. That image of the man standing there, looking at the wreck amidst which lie strewn the broken bodies of his loved ones. The wreck that he caused.

Sometimes there is no more hideous fate than to survive a wreck.

Another year has gone. Another year has crawled into its shell and lowered its burden upon my aching shoulders.

Sometimes I feel so tired, so very tired. There is no shoulder upon which I can lay my weary head and rest for a while.

(31st December 1976)

* * * * *

An extract from a letter to Yawar begun on 30th December 1976:

But the real problem of communication is at the personal level. Intellectual communication is a great necessity, but one can manage without it (there are substitutes – books, for example). What one cannot manage without is this human communion, this being able to relate directly as one person to another.

This world is an awfully lonely place. Each of us, encased in his hard shell, travels along his own solitary path, his only friend himself, all others his foes. What one desperately wants is what is hardest of all to find: someone to care for oneself not because of what one is, or what one has, or what one can do, but just because one is. And for such a person one would feel the same way. It is a free gift of nature that when one is young one finds several such persons in one's immediate family: parents and siblings, and to a lesser degree some among the next circle.

It is not a gift to be despised, though we tend, as we grow up, to first take it for granted, and then neglect it. Yet these are relationships that, if mutually nurtured (the enmeshment has to come from both sides) and if not abused (one person using the relationship to exploit the other), can last a lifetime, and provide something extraordinarily rare and practically priceless. The alleviation of human aloneness, a breaking of the walls that cut off one person from every other, a reaching out and touching, a finding of joy and pain in another's joy and pain, that thing, in short, whose only correct name is love, even though that word is so often misused and trivialized. This rare thing is there for the having. And the keeping.

As a young person grows into adulthood he often falls in love. If this is more than sexual infatuation it too is love of the same (i.e, the real) kind, though much more intense. But (perhaps because of this very intensity) it does not, it cannot, last very long. If it has led to marriage then soon, all too soon, the two find confronting them the same, age-old, universal problem, that of establishing true communication. This takes time, and during this long haul only your original family cares for you just because of who you are. It is only with them that you can have this authentic communication. To abandon them, to cut off all links with them, is to launch oneself alone on a vast, empty, hostile ocean, trusting to chance for some lucky landfall where one might find others to love and be loved by in return.

* * * * *

If wishes were horses – guess who'd get trampled?

– *Afroze (in a letter, Jan 1977)*

* * * * *

The true purpose of human life is the service of an absent God.

Because he is missing, this service necessarily involves restoring him to our world by acting in his place.

* * * * *

In spite of my poor circumstances and prospects sometimes a great surge of joy courses through my heart.

Why? Is it just the animal joy of being alive? The implicit realization that, however bad my lot, it could have been much worse? Some secret spring of unreasoning hope?

I do not know why it should be so, but it comes off and on. A surge as palpable as a rush of blood. Of joy – no other word describes it more aptly.

* * * * *

These days I am being publicly remembered. Twice in the past month we have been referred to in broadcast speeches.

It's a curious sensation. The feeling one would get if, looking at a cobra in its glass case, it were to rear up and flare its hood at you.

Except that I am in the glass case while the cobra is outside. The cobra also has the key.

* * * * *

30th March 1977 – another anniversary of sorts. Four years have gone by. Four years out of the not too many left. They have been hard, but I am immeasurably glad I have been able to put them to some use, not let them go waste. Still, they have been four desolate years, grinding me down. How many more? How many more can I endure?

* * * * *

Truly innocent are the poor and down-trodden. Whatever of evil they may have in their hearts, it is stifled by their helplessness, by the iron bands of their circumstances.

It is we, the better-off, the upper layers of society, who are the essentially guilty, however good or pure we may be. Either we oppress them ourselves, or we support the system which ensures their subjugation and deprivation, or, at the very least, we acquiesce in it. In any case, whatever good we have in life is at their cost.

Sometimes (often?) the only way of expiating this guilt is to sacrifice everything and join the innocents in their destitution, their want, their absence of choices, their defencelessness, their helplessness.

* * * * *

A record playing somewhere, somewhere far away. A woman's voice, faint but pure and clear, rises in song, rises high, soaring, above the stage scenery within which we live, above art and artifice. It is the voice of Woman, singing the sorrow of the world, the sadness of life.

It is as if I am sitting under a scraggy tree, looking out upon a blank, barren, parched landscape, beaten flat by the hammers of the sun. Scattered about are other trees, small, blackened, twisted and gnarled – but still surviving. Their roots grimly clutch the waterless soil, their thin, bare branches writhing in the desperate struggle to live on.

Over this desolation soars that pure, clear voice, pouring out the sorrow and sadness of the human heart. In it is a strange, elemental beauty, a beauty beyond human intent and effort, beyond even what nature can show; it is the unmatchable beauty that only the human heart sometimes creates. The same beauty one sees in the face of a woman riven with the pain of love.

It is this beauty that precludes despair, however deeply one is pierced by the poignancy of that voice. After the tears have dried one can still get up and trudge on through the wilderness.

* * * * *

The saddest of words are not "It might have been", but rather: *I could have done. Or, not done.*

* * * * *

When children most need the interest and attention and company of their parents, the latter are so involved in their own affairs they do not have enough time to devote to them. And when the parents most need the company and companionship of their children, the children, now grown up, are too busy with their own lives to spare much time and attention for them.

Is this inevitable? Or is there something wrong with the modern way of life?

Whatever the cause, it is a sad thing.

* * * * *

My unconscious has walked out of prison!

Some months ago, reading Solzhenitsyn's "The First Circle", I was struck by his portrayal of prisoners whose acceptance of their state had sunk so deep that even in their dreams they were prisoners. I realized this had happened to me, too.

Before my imprisonment I was very seldom aware of dreaming. But some time after it began I became aware of long, complicated, vivid dreams. Every night. They were seldom about prison life; in fact in most of them I used to wander all over the place, meeting all sorts of different people. Sometime later I realized that in these dreams, wherever I might be, I also always carried around with me, like a heavy stone, the awareness that I was a prisoner and would soon have to return to my prison. In every dream. Every night.

Bhutto fell on the 5th of July. It was something I had not dared hope for. My initial reaction was relief. No wild joy, no exultation, just a sudden relaxing of a constant tension, and a feeling of lightness. This relief lasted only a couple of days. Then Zia's announcement about holding a re-election in October raised the grim possibility that Bhutto might be returned to power. This thought worried me deeply; having scented the possibility of the end of my ordeal, it was exceptionally painful to contemplate losing it again.

It was during this period, while I was greatly upset and worried, that I realized that a significant change had occurred in my dream-life: I was no longer a prisoner in them! While, consciously, I was wrestling with the alarming possibility that this 'miracle' might be just a damp squib, that there might still be no salvation for me, my unconscious, apparently with a mind of its own, had decided I was, for all practical purposes, out of prison. Not that my dreams all became happy and joyful, but just that in them I was free. In fact, one of them was the saddest I can recall in a long, long time.

In it, I am standing in Faisal Square in Lahore; seeing for the first time the monument built there while I was in prison. I look up at it, a huge structure, something like one of those Aztec (or is it Mayan) pyramids. On top is a very big statue in bronze of a man, seated; I cannot make out who it is. I then go into the PIA booking office nearby. On its roof is a visible sign of Martial

Law – rifle-toting sentries in olive-green. Inside I make a booking to travel somewhere. At a nearby counter my wife is doing the same. As usually happens in my dreams, I do not see her; I am only aware of her as a presence.

In order not to cheat what Nabokov once referred to as "the delegation from Vienna" I should record that the girl at the counter is wearing a sleeveless dress and, pretending to check the entries she is making on the ticket, I rub up against her bare arm. She is apparently pleased, but pretends not to notice. (On second thoughts, this will probably disappoint the gentlemen from Vienna. They prefer elegant symbolism, not crude realism).

I leave the office and see Abbasi[8] stepping out of an army command vehicle parked outside. He walks towards the building, but I avoid him. Then I see a group of people, 20 or 30 of them, moving on the road towards the Assembly Chambers. A mixed group, men, women, children. Most have hair of a light, dirty brown colour (Namgay Doola?). Around the group is a cordon of policemen carrying staves; they are hustling and pushing the group forward. I realize that these people are Indian Christians who had sought refuge in Pakistan. They have been handed over to the Indian police who are taking them back.

The group is going away from me. At their rear is a young girl, her long straw-coloured hair falling down her back (the young Afroze?). As the policemen shout and push people forward, she is saying something out loud. It is in some strange tongue I cannot understand, but I can clearly feel the heavy charge of pain and sorrow that weighs down her voice. Then, without understanding her language, I understand what she is

---

[8] SRM Abbasi, a son of the Nawab of Bahawalpur, who passed out of PMA with me. He retired as a Lt Gen and Zia made him governor of a province.

saying. And what she is saying is, "All we want is peace. All we want is to live in peace." Over and over again. There is nothing I can do for them. I put my head against the railing I am holding, and cry, and cry.

Soon after, I woke and it was time to get up. The dream was still vivid. Even awake, I could hear that voice sounding in my ears. And I could not restrain my tears.

One reason for recording this dream is that this phase of vivid dreaming may be coming to an end. A change is unfolding; I am becoming less and less aware of dreaming at night. In fact these days (end of July) I tend to vaguely remember that I did have some dream the previous night but I cannot recall its details.

Another noticeable change: my stomach acidity problem has improved considerably for the first time since the spring of 1972.

Consciously, while I have some grounds for hope, I am still not prepared to transform hope into expectation. My unconscious seems to have made its own decision and has walked out on me, leaving me sitting here alone in my prison cell.

* * * * *

A person in love, any person, often behaves towards the beloved as an angel would. Does this not show that in every human being there is the potential to be an angel? If each of us loved everyone else, would we not all become angels, and this world a heaven on earth? And is this not what the Heaven of the Hereafter will be – a place where each loves all with the same intensity and selflessness as a lover does the beloved?

* * * * *

Prison has made me rather paranoid. I'm like the blindfolded man in the middle of a room who is, off and on, suddenly struck unheralded blows. After a while every whisper and furtive footfall, every secret scrape and rustle, becomes charged with menace; preliminaries to another blow that *they* are about to deliver.

So it is with me. Anything out of the ordinary routine – a delayed letter or newspaper, some perceived change in an official's attitude – sets my ganglions jangling and I begin to suspect some fresh indignity or injury is being plotted against me. Such premonitions have been right just often enough to prevent me from getting out of this mindset.

* * * * *

Epitaph for Nomi

(A letter to Shahzad and Nighat, whose son had been tragically stricken with cancer in his young teens, and had just died)

I have no words to describe the deep sorrow I feel at this great tragedy.

There is little that anyone can do or say to alleviate your grief and anguish at this terrible loss. Still, I wish I could somehow convey to you my own sense of absolute certainty in the reality of a life beyond this life in which all partings are ended, all wounds healed, all wrongs righted. Seen in this perspective the earthly disasters that befall us lose some of their cutting edge.

It might also help a little for you to be aware that, though no one can really share in your heartbreak, many other hearts are also stricken with a deep and abiding ache at this calamity. It sometimes happens, though rarely, that just to know a person is to love them. He was such a one for me.

* * * * *

There is such extraordinary beauty in the very young – babies and small children, the young of animals. Perhaps it is the beauty of life itself – life before it has been sullied, before it has been deformed by the ugliness attendant upon living in an imperfect world.

But oh! the sheer joy of seeing a little kitten at play.

* * * * *

A heart of oak. My father had one. Life treated him pretty badly. One after the other, most of his hopes and ambitions, both public and private, were blasted. And yet we who were near to him never saw him falter or withdraw, or be downcast and disheartened, or seek solace in bitterness and meanness. Bitterness there was, but it did not warp his essential nature, his basic kindness and gentleness, his love and affection for his near ones, his capacity for pleasure in the ordinary things of a simple life. The bitterness must have been due mostly to the unjustness of it all. He was a good man; good at his work; good to other people. And yet an evil fate seemed to dog his footsteps throughout his life. He took it magnificently: baring his chest and receiving without flinching every outrageous sling and arrow. It was a superb demonstration of how the greatness of the human spirit can render meaningless the vagaries of an adverse fate.

It was not that his heart was hard; on the contrary he was very soft-hearted. It was just very strong. Of course, it gave out in the end, but so long as it beat it was a heart of oak.

My brother, too, had a very strong heart, though in a different way. When the truck smashed into his little MG roadster, a portion of his breastbone pierced his heart. And yet it beat on for the better part of a day. They only found out at the autopsy.

I hope I have also inherited a strong heart.

* * * * *

13

Lahore

(March 1978 to Sept. 1979)

Early in March I was moved to Kot Lakhpat Jail, situated a few miles outside Lahore. This was the first prison in which I found myself in accommodation specially built to house 'A class' prisoners. I had a fairly large bedroom, reasonably well furnished, with an attached bathroom which had a wash basin with a tap, a shower and a flush toilet. After the facilities in previous prisons this was sheer luxury. There was a verandah and a small yard outside the room. The suite had proper doors and windows, the only bars being on the yard gate.

Even more pleasing was that I found Asif Shafi, my Attock comrade, there. We had a happy reunion, and he helped me to settle in, even undertaking to train the prisoner-cook provided to me. However, after about a week, his release orders came through. Even though I knew I would miss his company, I was very happy to see him set free.

On 18th March the full bench of the Lahore High Court found ZA Bhutto guilty and sentenced him to death. A couple of days later I learnt he had arrived in Kot Lakhpat Jail and was being kept in a death row cell. The wheel of fate had come full circle. He had tried to send me to the gallows, and now he was facing that same prospect. We were both in the same prison, housed a few hundred yards apart, except that I was in an 'A class' suite whereas he was in a small death row cell with an iron-barred door and a squatting toilet. The prison doctor visited me every few days on his routine rounds; before coming to me he would visit Bhutto and often talked about him. It seemed he wasn't taking his incarceration very well. The doctor told me that Bhutto obviously didn't like the solitude and would try and keep him engaged in conversation for as long as possible, plying him with tea from his flask. One day a fly fell into Bhutto's tea; he threw it away and said: *Doctor sahib, death is better than this kind of life.*

Jamila could now visit me every week. The jail offices had a proper visiting room, unlike previous ones where we had to meet in the Deputy Superintendent's office. She stayed for the better part of the day and we would share the lunch she brought. A couple of times when coming to meet her I crossed paths with the Deputy escorting Benazir Bhutto to meet her father in his death row cell; the Deputy and I would exchange greetings and he told me later that on the first occasion she wanted to know who I was. She probably told her father.

While Zia had successfully resisted the urgings of the politicians and media to release us, he was also getting pressure from another source. Three officers sentenced by the Air Force court-martial were also still in prison, and the air chief, Air Chief Marshal Zulfiqar Ali Khan, was keen to release them. When he told Zia about this the latter asked him to hold on so that he could coordinate the release of the army prisoners with theirs. When nothing further happened Zulfiqar raised the issue with Zia again, and was again asked to wait. Zulfiqar finally got fed up and told Zia he was releasing the air force prisoners in three days, and if he wished he could do the same. Obviously, Zia had no such intention, so in April Sq Ldr Ghaus, Wing Comd Hashmi and Gp Capt Sikandar were released, while the three of us (Aleem Afridi, Farouk Adam Khan and I) remained incarcerated.

I had of course long ago given up all hope of Zia letting us out, and had turned to the courts as the only way out. As mentioned earlier, in December 1975 I had submitted a review petition to the Supreme Court regarding their judgement in our case. This had finally been admitted in March 1976, and in November 1977 I had sent them a request to set an early date for hearing the petition, but had heard nothing further. On April 9, I repeated my request.

I had also spent a good deal of time searching for other useful legal openings. Finally I discovered a possible ground for striking down my conviction, namely, that when I was arrested, tried and convicted for "seducing or attempting to seduce a member of the Armed Forces from his allegiance to the Government", there was no legally constituted government in Pakistan to which the military owed any allegiance. The basis for this contention was that Yahya Khan had transferred his office to ZA Bhutto and, since the former had been declared by the courts to be a usurper, Bhutto's government could not be regarded as legally established.

However, such a contention had to deal with the fact that the Supreme Court, using the notorious "doctrine of necessity", had in 1973 declared Bhutto's Interim Constitution of 1972 to be valid and legal. I developed quite a plausible argument against this conclusion; the difficulty would be to get the Supreme Court to adopt this view in the face of their earlier judgement. I laid out my thesis in a memorandum which Jamila took to former Chief Justice Muhammad Munir, an old family friend. After studying it he wrote on it: *In my opinion such a petition has reasonable chances.* She also took it to a senior lawyer, Dr Khalid Ranjha. He got quite excited about it and offered to work gratis on drafting the petition. Since I already had a review petition before the Supreme Court we decided to submit this as a *habeas corpus* writ petition to the Lahore High Court. When it was ready I despatched it to them on May 11 through the jail authorities.

Meanwhile, in early May I was informed that my review petition would be taken up by the Supreme Court on May 13. As I learnt later, this prospect set off a remarkable chain of events. The court had also notified the government about the hearing, and this wound its way up the chain to the Attorney General, Sharifuddin Pirzada. When he looked at the review petition he concluded it was likely to be accepted by the court (it is also possible he was given a 'heads-up' to this effect by the court). Pirzada went to Zia and told him I was likely to be released by the court, and if he wanted to gain the credit for it he'd better act quickly. On May 11 the Ministry of Defence issued an order remitting the unexpired portion of my sentence and those of Aleem Afridi and Farouk Adam Khan, and told GHQ to ensure it was implemented immediately.

For me May 11 dawned as just another day in prison. In the morning I handed my writ petition to the Deputy Superintendent to forward to the High Court. After lunch, around 2 p.m, I got a small chit from the Deputy saying my release order was on its way and I should pack up! I asked for some more details and was told the prison had received a phone call from GHQ saying that an officer was flying down from Rawalpindi to Lahore with my release order and I had to be released by that evening. Quite surprised, I packed my belongings; few as they were, this was soon done. Then a big dust storm started and the electricity failed. Since it was still quite hot, I pulled a chair out on the verandah and sat there; I had to cover my nose and mouth to keep the dust out, but at least the wind was a bit cooling.

After waiting for a couple of hours I made enquiries and was told they were still waiting for the release order to arrive. After another fruitless wait I got quite angry; my capacity to handle practical jokes had obviously been seriously eroded. I sent word to the Deputy that I was staying put and not leaving that day whether they got the order or not. I had my things unpacked and my room restored to its original condition: bed made up, clothes hung up in the closet, books and writing things back on the desk, and my toilet things laid out in the bathroom.

Later I was told that Jamila had also learnt about my impending release and she, my brother Daniyal, his wife Salomi, and some other relations had arrived at the jail late that afternoon. The staff provided chairs and they sat outside the main gate for a couple of hours. When nothing happened they returned home and Jamila rang up a friend of ours in Rawalpindi, Gen Ihsan (retired by Gul Hasan, he had now been given a high position in the Ministry of Defence by Zia; later on Zia fired him when Ihsan made a joke in public that Zia thought was directed at him). Ihsan called back after a while and said the dust storm that had triggered the power outage and forced me to sit in my verandah had also closed down Lahore airport. The flight of the officer bringing the release order had been cancelled and the order would now arrive tomorrow.

Next day, around mid-morning, a warder brought a message that the release order had arrived and I should pack up. I sent him back saying I was not budging unless the Deputy himself came and told me that. Instead, he sent another jail officer to confirm that, yes indeed, the order was there and I should come to the office to be released as my family was waiting for me. I packed up and walked to the office where I signed the necessary papers and, with Jamila and Dani, walked out of the prison gates a free man again. It had been 5 years, one month and twelve days!

That day and next one are a blur in my memory. My mother's house, where we were living, was overrun with relations and friends continuously arriving as they learnt the news, bringing flowers and sweets; there was much noise and excitement. After spending the last four years in almost total solitude I was quite overwhelmed, and do not know how I coped. I do remember speaking to the children and other family in Canada on the phone. When are you coming, they asked?

The hearing of my review petition in the Supreme Court had been postponed from the 13th to the 15th. The court had appointed Mian Mahmud Ali Kasuri, one of the top lawyers in the country, as *amicus curiae* for the case, and I went to see him a day before the hearing. I tried to explain the arguments on which my petition was based but he did not show much interest; apparently my release had changed the significance of the case. Next day I appeared before a full bench of the Supreme Court presided over by Chief Justice Anwar ul Haq. He asked me if, in view of my release from prison, I still wished to pursue the case. I replied that I did, since my release had not reversed my conviction and sentence by the court-martial, which was the aim of the petition. The government was represented by a Deputy Attorney General who requested a postponement to enable him to seek instructions.

The Chief Justice suggested to the Deputy AG that, having released me, the government should perhaps actively consider quashing my conviction, to which the latter replied that if I put in an application to the government it would be sympathetically considered. Mian Mahmud Ali Kasuri also advised the DAG to this effect. I agreed to make such an application and the court then adjourned the case. A couple of days later I sent in the application. Months passed and nothing happened. I started pushing for the case to come up again, and this took place on June 23, 1979. The Deputy AG told the court that the government was still actively considering my application and requested that the case be again adjourned. This was accepted. Finally, on August 18, 1979 the Ministry of Defence informed me that my application had been rejected. By then I was getting ready to depart for Canada and it was too late for me to pursue the matter any further.

The same kind of charade was played out in respect of my writ petition in the Lahore High Court. It was heard by Justice MA Zullah on May 31, 1978. The Advocate General, Punjab, Aamer Raza Khan, appeared on behalf of the government, and also requested an adjournment as he believed that the government was actively considering annulling my conviction. His plea was accepted, and he then took me to his office and tried to persuade me to withdraw the petition since this would make it easier for the government to act on my case. I suggested to him that it was in the interests of the Zia government if my contention that Bhutto's government was not legally established was accepted by the courts but his only interest seemed to be to persuade me to abandon my case. He told me that Sharifuddin Pirzada, the Attorney General, had instigated my release, and he would also try and have my conviction annulled.

He offered to arrange for me to speak to Pirzada who would confirm this. His staff tracked Pirzada down in Karachi, got him to come to the phone, and Aamer Raza spoke to him and then handed the phone to me. Pirzada repeated the same line: drop the case and he would try and get my conviction annulled. Since, at the time, my Supreme Court case was causing some movement, I agreed to drop the High Court petition, and withdrew it a few days later.

There was nothing left for me in Pakistan. The children were already settled and studying in Canada, a country about which both Jamila and I had very happy memories from our two-year stay during the 1961-63 Staff College course. We decided that we would emigrate to Canada. I got the necessary forms and put in our application. This started the fairly long process which included a couple of interviews at the Canadian High Commission.

In early August, Asif Shafi and a couple of other Attock comrades met me and said that they were still being shadowed by intelligence personnel, and were having trouble finding jobs or other work because of this. They asked me to go and see Zia-ul-Haq and reassure him that there was no danger from us and this needless harassment should be stopped. While fully sympathising with their predicament I told them I was not prepared to ask Zia for a meeting. They suggested that, instead, I should just sign the visitor's book he kept at the gate of his residence, since they'd heard that he scrutinised this every day, and sometimes called people in to meet him. I agreed to do that during my impending visit to Rawalpindi.

Later that month, while in Pindi, I did just that. Sure enough, next day I got a call from Zia's Military Secretary asking me to come the next day to see his boss. Before the appointed time I drove to the army chief's house where Zia was still living. The Military Secretary seated me in the verandah as someone was with Zia. A short while later the visitor left and the MS ushered me into a large over-stuffed drawing room. After a short wait Zia walked in, dressed in a white *kurta pyjama.* I stood up and he shook hands with me and seated himself across the room. For a while we just looked at each other (I had decided I was not going to speak first). He then said: *I sent you to prison for doing something, and then I did it myself!* I replied with something about different circumstances. He asked me how I was, etc, and I then told him the reason for my coming to see him. He assured me that he had not given any such instructions, and did not see why it was being done; he would have this intelligence snooping stopped.

He then asked me what I proposed to do now. When I told him that I was planning to leave the country, he expostulated that I shouldn't do that, he needed people like me to help him put the country on the right track; I could choose whatever field I liked and he would let me run it. I demurred but, taken aback at his earnestness and insistence, I finally said I would think about it. He said he was going to be away for three days and I should come and see him after his return. That evening a friend asked me what Zia had said; I replied: *He offered me everything except his own job.* Later, when I told Gen Ihsan the same thing, he just laughed; obviously he knew Zia well.

I was not at all tempted to explore the genuineness of Zia's wide-ranging invitation since I was quite firm in my intention to go to Canada. However, I thought I would use this temporary access to put to him some suggestions on dealing with a couple of the most serious problems that the country faced; I had devoted some thought to these issues while in prison. These problems were the poor living conditions for the great majority of the people, and the lack of an institution that could ensure the unity of the country and check fissiparous tendencies. Accordingly, I spent the next couple of days writing a paper in which I argued that the solution to the first problem lay in improving the functioning of the district administration system, and suggested some relatively simple measures to achieve this. For the second one I proposed that the only institution capable of fulfilling this role currently was the Civil Service of Pakistan, and suggested what could be done to enable it to do so.

I sent the paper to Zia's Military Secretary. I was hoping for some response, but got none. It occurred to me that the MS may not have given the paper to Zia. To make sure I rang up Zia and left a message. Not hearing anything back, I tried again and this time was put through. Zia confirmed that he had got the paper but provided no reaction. That was the end of the matter, and I never heard from him again.

The process of obtaining our immigration visas to Canada took its usual slow course, but finally in August of 1979 they were ready. In September we said our goodbyes and left for our new home, Canada.

* * * * *

Looking back over the events of this decade it seems as if, on some sardonic whim, Fate had decided to entwine the thread of my life with those of Zulfikar Ali Bhutto and General Muhammad Zia-ul-Haq. My action after the December 1971 war that led to the removal of Gen Yahya Khan brought ZA Bhutto into power. It also had an immediate effect on the career of then Brig Zia-ul-Haq. He had recently returned from Jordan (where he had been a military adviser) under a cloud due to his involvement in the crushing of the PLO by King Hussein. Bhutto appointed Zia's friend and patron Gul Hassan as the Army chief, who in turn promptly promoted Zia to the rank of Maj.Gen.

During his political career Bhutto had twice seen the army remove the country's ruler, and he was acutely aware of this risk. Recognizing me as the mover in the last such episode he made sure I was quickly tossed out of the army. Then the Attock case provided him with the opportunity to not only deal with me permanently but also send a powerful message to anyone else in the army who might have such ideas. As a junior general Zia was picked to preside over our court-martial and Bhutto had him provide regular reports in person on its progress; these private sessions gave Zia an unparalleled opportunity to profess and prove his personal loyalty to Bhutto. At the conclusion of the trial Bhutto wanted me to be hanged, and Zia presumably assured him that it would be done; that is why the gallows at Campbellpur Jail was prepared and I was moved (along with Aleem Afridi) right next-door to it. I escaped the end they had intended for me by the narrowest of margins – one vote!

Having failed to get me hanged, Bhutto continued to pursue me with a vengeful ferocity. When he learnt that "life imprisonment" meant, in practice, 14 years behind bars, he had the rules changed so that such court-martial sentences really meant being locked up for life. He kept in his own hands all decisions regarding me, and made sure I spent nearly 4 1/2 years in solitary confinement far away from home. Over these years Zia assiduously built upon the foundation he had laid to convince Bhutto of his personal fealty. He succeeded so well that, when the time came, Bhutto picked him to be the next Army chief even though he was the most junior of the six contenders. And so it came to be that, when Bhutto over-reached once too often, Zia-ul-Haq was the one who removed him from power.

Knowing Bhutto, Zia was aware he could never sleep in peace so long as Bhutto was alive. There were many capital crimes that Bhutto could have been charged with; Zia picked the easiest one to prove, and in due course Bhutto was sentenced to death. Meanwhile, Zia had permitted me to be moved to Kot Lakhpat Jail, and it was there that Bhutto arrived as my neighbour, housed barely 100 yards away. We were both in solitary confinement, but he was in a death-row cell while I was in an A-class suite. When Zia could no longer keep me locked up, he released me, and it was as a free man that I heard, early on 4th April 1979, that Bhutto had been hanged the previous night. Not one of those other generals, any one of whom Bhutto could have picked instead of Zia as the army chief, possessed the cold-blooded ruthlessness needed to have him hanged. But it was Zia whom he picked.

But for his early promotion in 1972 due to my action, and the resulting opportunity provided by my court-martial to establish a personal relationship with Bhutto, Zia-ul-Haq would never have become Army chief. Nor, therefore, would he have become President of Pakistan. Perhaps then that plane in which he was flying on 17th August 1988 would not have inexplicably made a couple of strange loops and then dived straight down into the Bahawalpur desert.

I sit now in a faraway land, and it is but rarely that I view these events of the past unfold as if on a dim stage. Sometimes the side curtains move and it seems to me that in the shadows there I catch a glimpse of the grinning face of History's Black Jester.

How, in the end, it all turned out well (personally, at least)

The heroine of this tale: Jamila (1979)

Canada (1980)

Canada (1981)

Canada (1982)

Head of Nuclear Emergency Preparedness, Ontario (1983)

A reunion (1999): (*L to R*) Farouk Adam, Me, Saeed Akhtar Malik, Aleem Afridi, Iftikhar Adam, Nasrullah Khan

Toronto 2007: (*L to R*) Yawar, Wesley, Shawn (Afroze's children), Me, Afroze, Jamila

Postscript

Requiem for a Lost Country

[*I wrote this piece in August 2010 after the devastating floods that hit Pakistan that summer. It was published in the weblog* Sic Semper Tyrannis, *one of the few oases of reason and sanity in the wilderness that is public discourse in the USA these days. It belongs to my friend, Colonel W. Patrick Lang, a distinguished retired officer of the United States Army.*

It seems appropriate to reproduce it here as the conclusion of this memoir.

I should add, however, that the response of the people of Pakistan, especially the young, during the recent (2013) elections leaves me with at least a glimmer of hope for the future.]

The earth gods are not mocked lightly. They are slow to anger, but scorn them long enough, defy them long enough, and they will rise up and unleash their fury upon you. This is what they are now doing. This year their baleful gaze has focussed on hapless Pakistan. They could hardly have picked a better target: a country with limitless vulnerability and no defences. Over-populated, under-resourced, practically bankrupt, no governance to speak of, with a corrupt, bloodsucking ruling class, smothered under the weight of men with guns, some wearing uniforms, others sporting beards.

But it is not this country that I mourn. It will survive, as will its hardy people. Like the lowly of the earth everywhere, survival is about the only skill that their forbears bequeathed to them. They will pick themselves up, bury their dead, bind their wounds, and resume their threadbare lives – till the next calamity strikes. For, with Nature up in arms, it will be the drought next year, or the year after, or another flood, or some pestilence. If, at some point, the country begins to totter, one set or other of the hollow men with guns will take it over. No joy to be found there, for all they think about is war, all they care about is "the enemy". To them (even the well-intentioned ones) the country is just a base to be used to mount their unending campaigns.

The country I mourn is the Pakistan that was meant to be, that could have been. The Pakistan that its founder wanted, the country he conjured up in that vision he held out to the millions who laboured and struggled and sacrificed to help him bring it into being. A country founded on the ideals and values of Islam, in which all its citizens would be equal whatever their faith, with liberty, social justice and the rule of law for all.

Many forces opposed it, but the bitterest enemies were the self-proclaimed guardians of religion, for they knew their narrow, ossified creeds would have no place to thrive in the country that he wanted to build. He beat them back, as he did all the others arrayed against his mission. Muhammad Ali Jinnah's spirit was indomitable, but the arduous struggle wore out his frail body, and he did not live long enough to shape the country that he had created.

It is the loss of this country that remained just a dream that I mourn. This dream that is now dead. It survived the depredations of man, but with Nature now joining the assault, its time is finally gone. I mourn its passing as do many others: the dwindling numbers of those whose youth, like mine, was set ablaze by the promise of this vision, the others who came to it later but gave it their allegiance, all of us who kept the dream alive in our hearts and strove to bring it into being. And, above all, the few, the happy few, who stepped unhesitatingly forward when the dream beckoned that it may need their lives to ensure its survival.

It is necessary for all of us committed to this dream to bear witness to it while we are still around. Since, as it has faded, there have crept out of the shadows many who opposed its realization as well as those who now seek to desecrate its memory: those who claim that creating this country was a mistake, and the many others who falsely claim that it was meant to be a theocratic state.

We owe it to this shining dream of long ago, the Pakistan that could have been, and to ourselves, to step forward and say:

> *It was not inevitable that things had to be as they are. It could have been different. Our lives are proof that the dream was real, that it could have come about, that it was a goal worth striving for. Dreaming that dream and striving to bring it about is what has made our lives worthwhile, even if the world sees them as failures.*

(http://turcopolier.typepad.com/sic_semper_tyrannis/fb-ali/)

Appendix 3

The Attock Poems

[As I have recounted earlier, a day or so after my arrival at Attock Fort, when the initial shock of this sudden calamity had subsided, and I began to contemplate its implications for the future for me and my family, I was wracked by a deep despondency. On the surface I was coping and functioning fairly well, but deep down a terrible sadness sat like a dead weight upon my heart.

After two or three weeks they allowed us to buy some writing material to write letters home. I found in myself a great need to express what I was feeling by putting some of it down on paper. The only appropriate medium seemed to be verse; prose could not capture the hurt that was there. These are the poems I wrote in the succeeding days on the letter pad in my cell in the Fort.]

Fait Accompli

Behead a chicken,
See how it runs.

The plunging legs, still unaware,
Run from a danger dimly seen.
The flapping wings, not made for flight,
Fly from a flash unseen.
The foolish heart, faithful to the end,
Still pumps the useless blood
That gushes from the open neck.

Only the severed head,
The gaping beak, the glazed eyes,
Only they know there is no flight
From the accomplished fact.

When does a chicken die?
When the blade slashes, or
When the body drops?

Behead a chicken
And see how it runs.

The Lost Field

From far and near the vultures come
Onto this field of battle lost,
Swooping down with folded wings
To feed upon the fallen flesh.

It was a battle lost
Before the opening shot.
It was a legion doomed
Before its ranks were formed.
Besieged from without,
Betrayed from within,
Broken and shattered
The legion lies scattered
Upon this bloody plain.

And the vultures gather
To pick upon our bones.

Let us lie in peace,
For we are of the dead.
Take your tarnished victory,
Relish it while you can.

The Tumbrils are Here Again

The tumbrils are here again,
Creaking their mournful way
Through the hushed streets
To the murmuring square,
Where the hungry blade
Falls again and again,
Until it has consumed
Its daily feed of flesh.

The jackboots are here again,
Tramping through the nights,
Carrying off their prey
To some silent wall
Or the hissing chamber
And the burning pit.

From the rostrum shrieks
The same shrill voice,
Of Liberty and Fraternity,
And the Promised Land
That lies just beyond
The next pile of dead.

But in their shuttered homes,
The people only hear
The tramping of the jackboots
And the creaking of the tumbrils.

Freedom

Freedom is two butterflies chasing each other,
A flock of birds wheeling in the sky,
Trees swinging in the breeze.
Freedom is flowers blooming in the sun,
The sound of a child's voice,
Men joking together.
Freedom is a train's whistle in the distance.

Freedom is the glint of the sun
on the high hills around,
Freedom is the sound of traffic
beyond these walls.
Freedom is a man striding briskly
where his will takes him.

Freedom is a world
not forever quartered
by iron bars.

The Burden

Even the uncomplaining heart
Sometimes, in silent reproach,
Lays out the sorrows in the burden
It has carried over the years,
Making you drink once again
Each bitter cup of the draught
That has filled it over time.

Cold wet nights. A cold damp bed.
The flash of lightning, the clap of thunder,
Wind that would tear the roof away,
A child, alone and bewildered,
Taken from a home's warmth
Thrust among a crowd
Of rough and blustering boys
That spoke a strange tongue.

Dark nights of pain and anguish
A young boy keeps lonely vigils
Besides a pain-wracked mother
And prays and prays
That the pain might go away
That his life might be taken
In forfeit for the other.

A blaze of light, a world aglow
Beauty as never seen before
Love as never felt before.
A life transformed,
Filled with such grace and joy
As only the Garden could have known.
But soon, too soon, the casting out
Into the wilderness of separation
The long arid years of loneliness
The waste of heart and mind and spirit
Consumed in their agony of want.
So that at last, when the wandering ended,
It was only a shadow that came into the shade.

A Sunday morning, the rude awakening, a pencilled slip
News of a distant death, in a distant land
Yet it could not have been closer.
There was a strange hush over that house
When I arrived.
The pain in his eyes and voice,
The frightened face of a little boy,
The mother who had lost her mind.

That night he spoke of him and cried
The darkened room, the anguished voice,
The sound of tears
I carry with me to my grave.
He so young, yet old for his years
Bold and brilliant, fair and bright,
So full of life, and the joy of it.
The mocking smile, the impish eyes
They are no more.
Where is the abundance that sanctions such waste?

A hospital corridor, a passage to purgatory,
A woman's screams that will not cease,
A wayward child that will not be born.
Screams that lash a tortured soul,
Chained by love outside that door.

He was such a noble man, so gentle, so good.
(I speak as a son, but truly, unsentimentally)
No mean thought ever entered his mind
No cruel word ever fell from his lips
Tranquil of spirit, with a heart of oak
He walked his serene way, and would not stoop
To barter or bargain with the profane crowd.
But the pack of hungry mongrels
Would not let him be.
They harried and hounded him
Till they had driven him to bay,
But, though bloodied, he would not bow.
All the hurt they caused him
The worry and the grief
He hid in that great heart of his
Never in these times of troubles
Did we ever see him downcast
And in our home it was always
Sweetness and light.

But though they battered at his mighty heart,
It was, I fear, we who broke it.
He loved us so (it was a thing
You could almost touch,
A warm and lovely glow

That kept out cold and fear and hurt).
He loved us so, yet we deserted him,
I, stricken with love, he, stricken by death.
And even though I came back,
The damage must have been done.
So, at last, one day the tired heart gave out
He died, as he had lived,
Gently and peacefully.
It was I who brought the news,
And for a long while I stood
Outside the lighted window
Looking at the quiet happiness within,
That it was my lot to shatter.
Then I walked in.

A first-born child, much loved and cherished,
But in some tangled web enmeshed
Of contrariness and crossed purposes.
Misunderstood, misunderstanding,
A child of love, a lovely child,
Much loved and cherished
Growing up feeling unloved.

It was a mighty trumpet
That blew down Jericho's walls,
But it was the tinkle of a telephone
That brought my world tumbling down.
A world built up, brick by brick,
With long years of labour and love.
A life's handiwork, position, prospects,
Ease and security, even livelihood,
All wiped out in one cruel sweep.

And not for some foul crime,
But for a deed that
In another time, another place
Would have won honour and acclaim.

The next time it was the ring of the doorbell,
An angry buzz in the darkness before dawn,
And another world crashed into ruins.
The harsh rapping on the door,
The terror in a girl's scream,
The strange tramp of boots in quiet rooms,
Trampling life and love and happiness,
Past joys and future hopes.
A woman, bitter and defiant
As her world crumbled around her.
A girl shocked into sudden flippancy,
A young boy fighting to be a man.
And the stricken old lady who just turned
Her face towards the wall.

Searching through her cupboard, one of them
Came upon the box in which she kept
The letters I had once written her.
What, he asked, is in this?
Letters, she said, just old letters.
Something in her voice made him
Quickly turn away, though unaware
That his fumbling fingers were part
Of something that was perhaps
Ending a story those letters had begun.

No farewells, no tears of parting,
Only the bitter hardness
Of a useless rage.
The quiet of early morning,
The silence of a petrified city.

Then the weight of iron on wrist
The darkness of cloth on eye.
The swift flight from all that was good in life,
The blind entry into the abode of abandoned hope.
The bare cell, the iron door,
The clang of the bolt,
The click of the lock,
Then only the awful silence
Of the living dead,
For the bereaved mother,
For the widowed wife,
And the orphaned children.

Appendix 4

Understanding the Qur'ān's Message

The Islam that the Prophet Muhammad taught his followers was essentially a set of simple ideas and beliefs relating mainly to Allah, and the relationship of human beings to the deity. This overpowering vision transformed his followers, and led them to transform their world. The essence of the vision was awareness of, and belief in, Allah, the creator and sustainer of the universe, and acceptance of the role that Allah expected human beings to play in their world.

This original Islam had no dogma, no ritual, no elaborate code of laws, no special class of persons learned in the religion who guided and judged other believers; in short, none of the elaborate structure that now passes for Islam. This superstructure did not exist in the early centuries of Islam.[9]

Today's Religion of Islam (in each of its several forms) is a

[9] The raw material and the tools with which this superstructure was later gradually constructed (the standard Hadith collections and the doctrines of the *fiqh* schools) did not begin to be assembled and formalized until the third century A.H.

complex and rigid structure which overlays and obscures the original message the Prophet brought to humanity. Further complicating matters is the emergence of a class of self-styled religious authorities and "guardians", so that there is now a virtual priestly class in Islam where there was no place for one in its original version. From a simple but powerful ideology based on a few key concepts and beliefs, Islam has developed into an elaborate and complicated religion.

To rediscover the original essence of Islam we must seek it in the Qur'ān. All else that goes by the name of Islam today is superstructure added over the years by human beings. The original message that was named Islam, and conveyed to humanity through the Prophet Muhammad, is now to be found in its pure form only in the Qur'ān, and nowhere else.

Unfortunately, this endeavour has been successfully defeated for centuries by the religious dogma that the Qur'ān's message can only be understood through reference to Hadith and the interpretations and commentaries of earlier 'authorities' (even though such a contention is directly contradicted by the Qur'ān itself[10]). If we discard such preconceived notions and artificial embargoes, it is quite feasible to understand the Qur'ān's essential message directly, from the book itself.

Understanding the Qur'ān

The Qur'ān is a record of the divinely inspired[11] utterances of the Prophet Muhammad over a period of about 23 years; these were recorded and memorized at the time of occurrence and compiled in

---

[10] See, for example, Qur'ān 6:115; 10:37; 39:28; 41:3; 54:17.

[11] The most accurate translation of the term *wahy* is "inspiration", and not the commonly used "revelation", which is a term borrowed from other religions. The primary meaning of the root for *wahy* is a swift signal, conveying some meaning or giving some indication.

the Qur'ān, though not in chronological sequence.[12] Because of this, and the long period over which the text gradually came into being, the Qur'ān's message cannot be understood by reading it as one would any other book. To understand its real message today we have to adopt a different method.

The first step is to discover the correct meaning of the terms and concepts that it employs. To do this we need to take their meaning as it was understood in the Arabic of that time, which may not necessarily be the meaning developed for them later on. Since a term can have several different such meanings, in the case of an important or ambiguous term or concept, we should deduce the correct meaning or meanings the Qur'ān assigns to it by putting together and examining all its usages in the text.[13]

Secondly, since various subjects and issues are repeatedly dealt with at many different places in the Qur'ān, to discover the Qur'ān's position on any topic we must put together all the Qur'ānic references to it and then study the coherent picture that emerges. (It is the failure to do this that often results in Muslims with differing views on some subject claiming the Qur'ān's support for their particular position by quoting one or two verses as proof).

The third step is to deduce the overall ideology the Qur'ān teaches, within which its positions on all the major issues that it covers fit in a consistent, logical manner. This process often

---

[12] The placing of verses in the text was directed by the Prophet; this would have been necessary to enable the *suras* to be memorized in a standard format (also see Qur'ān 75:17). Thus, the mixing up of the chronological order was deliberate, perhaps to break for future readers any link to particular events or circumstances, and thus generalize the message.

[13] For an example of how this method can reveal meanings quite different from those commonly accepted, see the discussion on the term *fath* in the author's article, *Al-Hudaybiya: An Alternative Version* (The Muslim World, Jan. 1981: 47) [Reprinted in Uri Rubin, ed., *The Life of Muhammad* (Aldershot, UK: Ashgate Publishing, 1998)].

necessitates the revisiting of conclusions arrived at earlier in the second, and even the first, stage, to see if an appropriate modification (text-supported, of course) would enable them to fit better into the total picture.

However, merely understanding what the Qur'ān says on the different subjects it covers is not enough; we also need to understand their significance and relevance for us today. The Qur'ān deals with many different types of topics: some concern the fundamental issues affecting human beings, others deal with matters of contemporary daily life, while still others refer to other-worldly issues in metaphorical language, stories of earlier times, and allegories and parables. What should these various elements of the Qur'ān mean to us today? How do they apply to us? The answer to these questions is to be found in the Qur'ān's teaching on the system of divine guidance for humanity through *wahy*, which we shall consider next.

The discussion in this paper is based on a detailed analysis and study of the Qur'ān using this method of interpretation and understanding, and on the resulting view of the comprehensive system propounded by it. References to Qur'ānic passages are given below for many of the arguments made but these are mainly for purposes of illustration and are not the only basis for the positions adopted; these positions are based on the detailed study and the complete picture that emerges from it.

The System of *Wahy*

The Qur'ān says that, from time to time, a few select human minds have been inspired (in some fashion) with a consciousness of Reality and Truth.[14] This can be understood as a direct,

[14] The Qur'ān uses the terms *al-Kitab* and *al-Haqq* to represent these concepts in the context of *wahy*. Among many references in the Qur'ān, typical are 2:213; 3:78; 10:37; 22:54; 35:31.

intuitive comprehension of Allah, and of the relationship of the universe, including humans, to Allah. These persons are also imbued with an imperative urge to formulate this vision in words, and convey it to their fellow beings.[15] The Qur'ān refers to such individuals as messengers.[16]

However, these messengers are, and remain, human beings,[17] and the minds through which this vision is given form and expression are just human minds.[18] They are confined within the limits of their own specific knowledge as well as the general level of contemporary knowledge and thought. Thus, when one of these messengers puts into words this inspired consciousness of Reality and Truth, it is inevitably framed in contemporary terms (which also enables the message to be comprehensible and relevant to its immediate audience). Similarly, the application of this consciousness to practical affairs can only deal with the actual circumstances then prevailing.

What such a messenger conveys is a *contemporary* expression of Reality and Truth. While confined within the limits of prevailing knowledge and comprehension, and dealing with then existing circumstances, the details of the message conform completely to the

---

[15] The phenomenon can be understood as somewhat similar (though of far greater depth and complexity, and, of course, significance) to that of the genius inspired by a vision and impelled to express it in great art or music.

[16] The common usage in English of the term "prophet", which carries several different meanings, can be misleading. The Qur'ān uses the terms *rasul* and *nabi*, both of which mean a conveyor of a message or information.

[17] See, for example, Qur'ān 12:109; 14:11; 16:43; 17:93, 95; 21:7; 23:23-50; 25:20; 41:6.

[18] There are many indications of this in the Qur'ān, but the definitive verse is 22:52. (The term *shaitan* is used in the Qur'ān for negative impulses, attitudes, thoughts, and other failings of the human mind).

initiating consciousness, and the truths and realities comprehended by the latter underlie everything it contains. In other words, where the message deals with abstract matters (such as the reality of Allah, the divine system, the Hereafter) it does so in terms which can be understood by its immediate audience, but this is nevertheless an expression of the reality of these matters insofar as the human mind can grasp them. Where the message deals with practical injunctions, these relate to contemporary matters, but conform to the fundamental principles and values that should govern all human conduct, anywhere, anytime.

For its own time and place such a message is completely true and valid, and applicable in all its detail. But in places where circumstances differ materially, and even in the same area after the passage of time, the validity and applicability of the message becomes increasingly limited. The practical injunctions are no longer fully relevant since people's ways of living and their social structures have changed, while the descriptions of abstract matters may no longer satisfy since human knowledge and modes of thought have advanced.

In time, another messenger is inspired with *wahy*,[19] and conveys to his fellow-beings another message, which is an expression of the consciousness he has acquired of the *same* Reality and Truth, but one dealing with the issues of that time and place, and appropriate to the then prevailing level of human knowledge. Since this new message, though it differs in its details, is still based on the original truths and realities, it preserves and verifies them.[20] In turn, this message also becomes

---

[19] On the series of messengers, see, for example, Qur'ān 2:213; 10:47; 16:36; 23:44; 35:24.

[20] On the preservation and verification of earlier *wahy*, see, for example, Qur'ān 2:89, 97, 101; 4:47; 5:15, 46, 48; 6:93; 10:37; 35:31.

out-of-date, and is replaced by another one more pertinent to the evolving human situation.[21]

According to the Qur'ān, this system of *wahy* has gone on throughout human history, with numerous messengers arising in different times and places. However, it came to an end with the message inspired to the Prophet Muhammad in the 7th century A.D. (or the 1st century A.H.). The record of this *wahy* remains available to us in the Qur'ān. This message is of the same type as all earlier *wahy* in that its expression is in terms appropriate to its own time and place, and its practical injunctions relate to the circumstances then prevailing. But underlying these surface forms is the same Reality and Truth that have been the basis of all previous *wahy*. There is, however, one major difference between the Qur'ān and earlier messages: it is the last of the series and no more *wahy* will occur to take its place. This means that for today's Muslims, and for their succeeding generations, the fundamental truths and realities that have always been conveyed through *wahy* can be discovered only through the Qur'ān.

This aspect led some Muslim theologians to advance the view that every word of the Qur'ān is applicable for all time to come, and this proposition has become a dogma among most Muslims. This is unfortunate, since not only is it impossible to implement this in practice, but it also contradicts the Qur'ān's own teaching on the subject, as discussed above. What is valid and applicable for all time to come is not each word of the Qur'ān but the truths, realities, principles, values, concepts, etc. that lie behind, and are the basis of, these words. It is these that mirror the Reality and Truth (*al-Kitab* and *al-Haqq*) that were the initiating source of the Qur'ānic *wahy*, and not the verbal

---

[21] On the replacement of an earlier *wahy* by a later one, see, for example, Qur'ān 2:106; 5:44-48; 13:38-39; 16:101; 98:3.

constructions that had perforce to be formulated within the limits of contemporary comprehension and knowledge, and, for the most part, dealt with contemporary concerns in a manner appropriate to the immediate audience and their circumstances.

To take one example,[22] the Qur'ān makes a few references to the slaves that were part of society in its time, usually in the context of prescribing measures to ameliorate their condition or to urge their freeing. However, nowhere does it call unequivocally for the abolition of slavery. It is obvious that the Qur'ān, while disapproving of the institution, tolerated it in the then prevailing form and circumstances.[23] On the other hand, the principles and values underlying the Qur'ān (e.g., human freedom, the equality of all human beings) are totally opposed to slavery. Which, then, of these two opposite positions (one derived from certain of its words, the other from its underlying principles) should one regard as the Qur'ānic injunction valid for all time to come? The answer is obvious, and applies generally to the issue of which element or aspect of the message of the Qur'ān is valid for us today (and for succeeding generations).

The Qur'ān itself makes this explicit. There is a set of three passages[24] that introduce the term *umm al-Kitab* (the essence or core of the divine message), the only such usage of this term in the entire Qur'ān. Read together, these passages in effect say that every divinely inspired message had underlying fundamentals

---

[22] Other examples, to which the same argument applies, are the injunctions regarding the punishments for various offences, retaliation or *qisas* (2:178-179), polygamy, concubinage, women witnesses, etc.

[23] It was a relatively benign form of slavery, more akin to indentured service, and was an integral part of the social and economic systems. Slaves were lower status members of the families, clans and tribes that constituted Arab society. Most of them would have lost both livelihood and protection if suddenly freed.

[24] Qur'ān 13:38-39; 43:2-5; and 3:6. Linked to these is 39:23.

that endure permanently, while other parts, which related to its particular time and circumstances, were of temporary validity. These fundamentals are also contained in the Qur'ān, and they are the permanent parts of this divinely inspired message, with the rest of the book being similar to the transitory elements of the earlier messages (as those with knowledge and understanding can discern). In several other passages this basic concept of certain elements of the Qur'ān having deeper, lasting significance is expressed using the term *ahsana* (the best).[25]

What, then, are these fundamentals, the essence of the message that the Qur'ān brought into the world, which remain valid and applicable for us and for succeeding generations?[26] We can usefully consider this in three parts : the fundamental truths underlying the system within which we exist; the basic rule governing human conduct; and the permanent values which we should adopt and uphold. These are the foundations and the fundamentals of Islam.

The Quran's Fundamental Truths

These relate to the reality that underlies the universe, life and our humanity (issues such as Allah, Allah's interaction with human beings, the system of the universe, the role of human beings, human accountability). In a short treatment such as this we shall only highlight two of the most relevant and important ones : Allah, and the purpose of human life on earth.

---

[25] See, for example, Qur'ān 29:46; 39:18, 55; 92:6-7, 9-10. 7:145 applies the same concept to Moses' time. Also see 98:2-3 for the same meaning conveyed in a different form.

[26] Future generations will need to keep going back to the Qur'ān to see if, from their new perspectives, it provides any new guidance for their times and circumstances.

Allah : The primary message of the Qur'ān is the fact of Allah, and the need for human beings to believe in this fact, and thus accept Allah as the creator and sustainer of the universe and all that is in it, including them. This belief and acceptance, however, can only be as an act of free will.[27]

The starting point of the Qur'ān's discussion of this subject is that there is (and can be) no proof of Allah's existence; that is why it urges human beings to believe in it with such urgency and force, and at such considerable length. It urges us to study the universe around us, to look within ourselves, to ponder the consequences of our decision, and then, as free and rational beings, decide whether we should make the assumption that Allah exists, and believe in this. As part of this discourse the Qur'ān seeks to convey to us a concept of Allah through his attributes (the *asma al-husna*) and "actions".[28]

The Role of Human Beings : The Qur'ān puts it thus :

> "And when your Rabb said to the *mala'ikah*, 'I am going to place in the earth a *khalifah*.....' ".[29]

This passage (and similar ones[30] narrating the allegorical story of Adam) deals with a central theme of the Qur'ān: the emergence of human beings on earth, and the special position and role that they have in their world. The message that this

[27] See, for example, Qur'ān 2:256; 10:99; 13:31; 18:29; 32:13; 43:20; 74:55; 76:3.

[28] The concept of Allah given in the Qur'ān is one our minds can understand and relate to. It cannot convey the "reality" of Allah, since this is, by definition, beyond the limits of finite human understanding.

[29] Commenting on this verse (2:30) Ibn Khaldun wrote in his great treatise on human history : "This is the meaning of civilization" (*Al-Muqaddimah*, trans. F. Rosenthal (New York, 1958), vol. I, 85).

[30] 2:30-38; 7:11-25; 15:28-42; 17:61-64; 18:50; 20:115-126; 38:71-85.

allegory conveys is that we achieved our human status when our minds reached a certain stage of development, and in doing so acquired the potential to control and harness all the forces of nature. We were also endowed with free will, which enables us to use these great powers for any purpose we choose.

However, the Qur'ān reminds us that we developed these potentialities and powers so that we could perform the special role of *khalifah* that Allah intends for us in our world, a role that requires us to use them on his behalf. To do this we have to resist the constant temptation to use these capabilities for our own ends.[31] To assist human beings in this ongoing struggle, Allah has periodically sent us guidance through his messengers.

The basic meaning of *khalifah* is one who takes the place of another;[32] thus it is used to signify a surrogate or substitute,[33] and also a successor. The key component of the term's meaning is that the surrogate or successor functions in the other's place, and on his behalf. This is the special role the Qur'ān says human beings are capable of, and which we are offered: that of acting in Allah's place in our world.

The Qur'ān bases its doctrine of the *khilafat f'il ard* on, firstly, the proposition that Allah created our universe, and that it, and all things in it, exist and function according to his laws.[34] Beyond this, however, he has chosen not to intervene; he

---

[31] The Qur'ān uses the term *mala'ikah* to represent the forces of nature, and the term *Iblis* to represent those elements of human nature that tend to lead humans away from Allah's path.

[32] See, for example, Qur'ān 7:69, 74, 142; 11:57; 19:59; 25:62.

[33] "*An inscription from South Arabia (in a language cognate with classical Arabic) shows that the corresponding word was there used about AD 543 in the sense of 'viceroy' and this usage may have affected that in classical Arabic....*" (W. Montgomery Watt, *Islamic Political Thought* (Edinburgh: The University Press, 1968), 33). This usage is confirmed in Qur'ān 7:142.

[34] See, for example, Qur'ān 13:15; 16:49; 21:33; 30:25-26; 65:12.

withholds himself from acting in this world as a causative agent, and does not interfere in the operation of its natural laws.

If we pause and think about this for a minute, we can see that this, in fact, is how things actually are in our world.[35] Everything in the universe is bound in a certain mode of existence and behaviour (with one notable exception). The mighty galaxies travel in pre-determined paths and speeds; the huge stars follow ordained life cycles; the planets move unwaveringly in their orbits. The microcosm is no less firmly regulated than the macrocosm; each particle has its inherent properties and must conform to them and the laws that govern them; the sub-atomic universe appears to be as unitary a system as the wider universe. Life itself comes into being and develops according to evolutionary laws. Plants live and die in the established rotation of the seasons. Animals exist bound in the iron bands of instinct, their behaviour fixed within narrow limits.[36]

Nowhere in this orderly universe do we see a "divine hand" intervening to alter the operation of the system. As human knowledge has increased, previously inexplicable and seemingly random natural events, which were once ascribed to God's doing, can no longer be so considered. It is now known[37] that

---

[35] This, surely, is the primary test for the validity of a belief : that it is consistent with our observation and experience of the real world. It is to avoid this test that religions (including the Religion of Islam) seek to distract and over-awe their followers through their otherworldly and supernatural dogmas and beliefs, which they are required to blindly accept.

[36] Some recent research indicates that higher mammals may display some rudimentary volition. This would be entirely consistent with the evolution of the animal brain into the human form.

[37] Of course, many still don't know this, or refuse to acknowledge it. After every natural disaster some religious figure or other will invariably claim that this was God's punishment on sinners. They (*continued on next page*)

they all have natural causes, and occur according to natural laws.[38] The whole edifice of science and technology is based on this reality.

The second basis of the doctrine of the *khilafat f'il ard* is that the only entity in the world with complete freedom to act in any way it chooses is the human being; we have total freedom of choice, so much so that we even have the choice of refusing to believe in or acknowledge Allah[39]. Here again, as in the natural world, Allah does not intervene in or direct human affairs, nor does he affect the outcome of human actions and choices (even the Prophet is told in the Qur'ān[40] that there is no guarantee that he will see the defeat of those opposing him, or that he will not die before the success of his mission. What could be more definitive than this?).

This is also what we actually observe. Human beings, in contrast to everything else, appear free to order their lives and determine their behaviour in any way they choose. Recorded history and our own observation disclose human beings living, individually and collectively, in many different modes, according to the highest standards conceivable as well as the lowest, in pursuit of all kinds of aims and goals as well as none at all, performing actions which we would call saintly, and also those of the utmost depravity. This vast variety is proof enough that neither our inner nature nor any external constraint compels

---

blissfully ignore what kind of God it would be (on whose behalf they so smugly claim to speak) whose wrath fell mainly on the poor, the helpless, and innocent children.

[38] We should remember that chance and randomness are part of the natural system, as the Qur'ān recognises. See, for example, 3:139; 7:94-95; 10:24; 11:9-11; 21:35; 30:36-37; 53:43; 89:15-16.

[39] See, for example, Qur'ān 10:108; 13:31; 18:29; 43:40; 74:55; 76:3.

[40] See 10:46; 13:40; 40:77; 43:41-42.

human beings to live and act in any particular pattern or mode. How they live or what they do is for them to decide. It is true that not many of us are able, in practice, to make such free choices, but, in principle, there is no insurmountable barrier to prevent us from doing so. What any human has done, it is possible for other humans to also do.

We also see that the outcomes of human actions and choices follow no discernible pattern based on their ethical or moral quality, or their conformity or otherwise to any divine or religious directives. However, influenced by religious teaching, many people still consider occurrences affecting humans and human societies to be due to specific decisions by God. When, as often happens, good deeds have bad outcomes while evil actions result in gains, the wicked prosper and the virtuous or innocent suffer, such believers are forced to resort to implausible excuses and arguments, ignoring the implications for the kind of God who would act thus, and for the omnipotence and justice of the divine system they claim is operating in human affairs. Such a claim is supported neither by the Qur'ān nor by our own experience.

The third basis of this doctrine is that Allah acts in our world through the agency of humans.[41]

Based on these premises the Qur'ān expounds the doctrine of the *khilafat f'il ard.* Affirming that Allah withholds himself from acting as a causative agent in our world, and acknowledging the freedom of choice and action that humans possess, the Qur'ān offers human beings the role of Allah's *khulafa f'il ard.* It urges us to assume the responsibility of acting as Allah's surrogates, to act in his place in our world, to work towards his goals, to do all that he would have done in our world if he had so chosen. Apart

[41] See, for example, Qur'ān 2:251; 4:75; 22:40; 36:47; 47:7.

from the pivotal allegory of Adam, the Qur'ān refers to this task, this role, this Allah-human relationship, repeatedly and in many different ways.[42]

The *khilafat f'il ard* is indeed a heavy burden for us to assume. In accepting it we agree to live our lives for Allah, and not for ourselves. Whatever obligations he has assumed in our world, we undertake to fulfil. (For example, when the Qur'ān says (6:11) that the provision of sustenance to every living creature is Allah's responsibility, this is an obligation that we are expected to discharge on his behalf). To be able to fulfil all these responsibilities we have to first develop the great potentialities with which we have been endowed. With these we must strive, individually and collectively, to further his goals, not our own. We stand for Allah in our world, and must fashion ourselves and our actions accordingly.[43]

Basis of Conduct

The Qur'ān uses the term *amal as-salihat* to prescribe the essential basis of human conduct. The primary meaning of the root for *salih* is to remove a shortcoming, defect or handicap, and bring something to its proper and rightful state. In the context of the *khilafat f'il ard,*[44] this basic rule requires us to make good our deficiencies and limitations, internal and external, in ourselves and others, and achieve the stature and

[42] Some of the other concepts the Qur'ān uses to expound this doctrine are *ahd Allah, auliya Allah, ansar Allah, shuhada l'illah, amana b'illah, dhikr Allah, sabil Allah,* and still others. It is not possible to discuss these in this brief treatment.

[43] We have forgotten the real significance of the tradition followed from the earliest days of Islam whereby a Muslim commences every action with a *b'ismi'llah* : in the name of Allah.

[44] See, for example, Qur'ān 24:55.

capability necessary for beings who must act on behalf of Allah in the world. It requires us to act, individually and collectively, to bring this world into its rightful condition, and thus, through human instrumentality, to indirectly restore Allah as a causative agent to a world within which he has chosen not to act as one.

The overriding importance that the Qur'ān attaches to this rule of action is shown by its linking of this conduct with the essential requirement to being a Muslim: *iman* (belief in Allah), and the frequent use of this formula as the conduct for which Allah promises the ultimate reward.[45]

Permanent Values

The values that should permeate our lives and govern our actions can be derived from the Qur'ān, many of them from the attributes of Allah (of course, not all the attributes are applicable to human beings). The most important of these are :

<u>Freedom</u> **:** This, according to the Qur'ān, is the essential and distinguishing quality of human beings. Thus, maintaining our humanity requires us to promote and preserve this value for each and every one of us.

<u>Love</u> **:** The love that the Qur'ān extols is the one based on the term *rahma*. The root for this term is a name for the womb, and it refers primarily to the nurturing, compassionate love that a mother has for her child. Maternal love (and the response it elicits in every human being) is the foundation for our humanity (it may well be the key factor that propelled us to evolve to our present state). Perhaps the most frequently used names for Allah

---

[45] There are at least 25 references relating to the reward of *al-jannah* (e.g., 2:25; 18:30; 22:14; 32:19; 45:30; 84:25) and another 24 or more speaking of reward generally (e.g., 2:277; 3:56; 35:7; 41:8; 103:3).

in the Qur'ān are *ar-Rahman* and *ar-Rahim*, signifying the nurturing love that the Creator has for his creation. (Unfortunately, when Islam became a religion, the conception of Allah was derived from the most powerful being then on earth, the absolute ruler who wielded his power according to his whims, unconstrained by any rule or law or other consideration. For his helpless subjects, the highest and most desirable virtue in such a ruler was mercy, and so that is what *rahma* came to mean, and so it has remained).

Justice : In the Qur'ān's references to Allah's dealings with us the constant theme is that of justice. This value should figure prominently in our dealings with each other, and in the societies we establish.

Conclusion

Muslim societies everywhere are in crisis these days. The hopes aroused in the last century at the dawn of their freedom from colonial bondage have been dashed. One after another, for one reason or another, their attempts at modernization and advancement through adopting secular ideologies and systems foundered. The inevitable backlash brought with it a resurgence of the Religion of Islam, with its proponents seeking a return to its rigorous implementation as the only solution. Other proponents have ascribed the failures to foreign conspiracies and aggressions, and have taken up arms to wage Jihad against the "enemies of Islam", while enforcing religious practice and law among believers under their sway.

Recently, a new generation of young Muslims has come forward, demanding freedom, democracy and justice, and seeking to take their place in the modern world. Even though they acknowledge allegiance to Islam, their aims and ideas set them on a collision course with those who seek to impose the

Religion of Islam over all aspects of life in Muslim societies. At some stage this new generation of Muslims will come face to face with the awful dilemma of having to choose between the values and principles that motivate them and the Religion of Islam, as defined and established by its guardians. And so a new and dangerous fissure will open up in the Muslim world, further imperilling its prospects of emerging from the dire straits it is in.

This stark choice can be avoided if it is realised that this religion that calls itself Islam is very different from the original message of Islam. That message is still available to us in the Qur'ān, but only if we try to understand it from the book itself rather than through all the intermediaries and interpretations the Religion of Islam interposes between us and the original ideology of Islam that the Prophet Muhammad brought to humanity. Only in this way can educated, forward-looking young Muslims assume their rightful place as the vanguard of a revival of their faith and their societies, instead of either succumbing to a sterile orthodoxy they secretly reject or becoming renegades and outcasts.

For Muslim societies throughout the world it is the only way out of their current tragic state and the grim future that otherwise looms ahead.

